Yskak Nabi

Principles of Descriptive Geometry

AF209873

Yskak Nabi

Principles of Descriptive Geometry

Textbook for students of technical speciality

LAP LAMBERT Academic Publishing

Publisher:
LAP LAMBERT Academic Publishing
is a trademark of
Dodo Books Indian Ocean Ltd. and OmniScriptum S.R.L publishing group

120 High Road, East Finchley, London, N2 9ED, United Kingdom
Str. Armeneasca 28/1, office 1, Chisinau MD-2012, Republic of Moldova, Europe
Managing Directors: Ieva Konstantinova, Victoria Ursu
info@omniscriptum.com

Printed at: see last page
ISBN: 978-3-659-35169-3

Table of contents

1

Chapter 1. METHOD OF PROJECTIONS

1.1 Essence of projections method

The operation of projection lies on a basis of construction of various images. This operation is performed as follows: an arbitrary point S of the space is taken as a *centre of projections* (Fig. 1.1) and a plane π' not passing through the point S, – as plane of projections. In order the projection of point A of space on the plane π' to construct, line SA is drawn through center of projections S to the intersection with a plane π'. Point A' is called *a central projection* of the point A, SA is *projecting line*. The set of projections   of points of a geometrical figure is called its projection. However, a straight line is defined with two points, therefore for constructing its projection the projections are constructed of the two points lying on it. As the projection of the straight line is straight line, this line is drawn through projections of these two points.

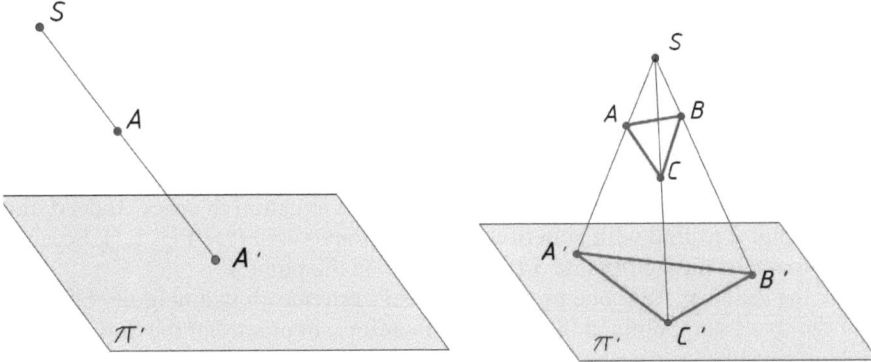

Fig.1.1   Fig.1.2

To construct of the projection of the triangle, it is necessary to construct the projections of its vertices and to connect them (Fig. 1.2). The central projection l' of curve l consists of the central projections of several points (Fig. 1.3).

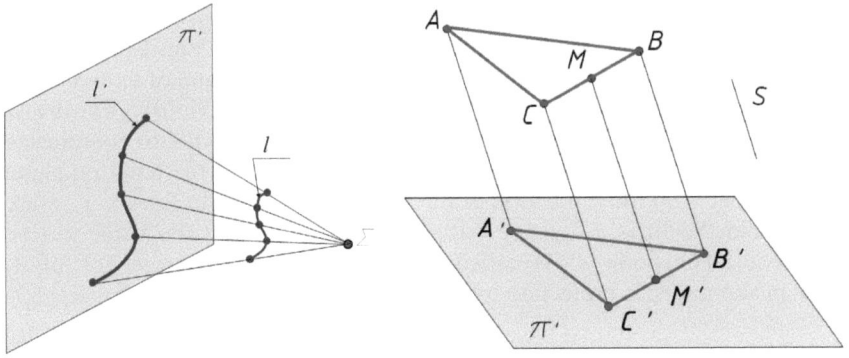

Fig.1.3   Fig. 1.4

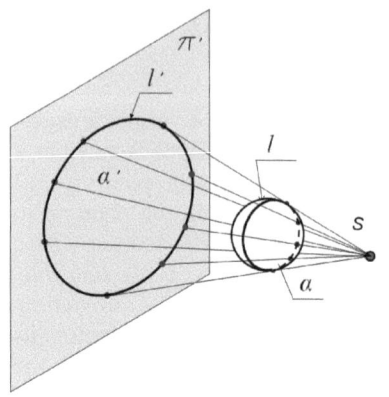

If the center of projections is located in infinity, projection lines are parallel to each other, and received projections is called *parallel projections*. Fig. 1.4 shows the parallel projection of the triangle *ABC*. Naturally, for construct the projection of the triangle, all its points to project no need. You can find the projections of points *A, B, C* - vertices of the triangle, and connect the points *A', B', C'*.

To construct the projection of surface, the projecting lines must be draw tangency to the surface (Fig. 1.5).

Geometrical place (a locus) of points of the tangent is called *a line of contour* (the picture is a line *l*), projection *l'* of line *l* is called an *outline* of surface.

Fig.1.5

1.2 Reversibility of drawing

If a form of an imaged object and its situation in space is determined with help of a drawing, drawing refers to the reversible drawings. You can conclude considering the methods of central and parallel projection shown in the figures 1.1 ... 1.5, that the images on the projections planes are irreversible, because the one central or parallel projection of the point doesn't determine its situation in space. Indeed, the situation of point *A* to find with help of the point *A'* impossible (see Fig. 1.1), because an arbitrary point of projecting line *SA* is projected on the point *A'*.

So, the drawing from one projection doesn't provide an unambiguity between points of the space and points in the plane π ', therefore, in order that the drawing was reversible, it must be complemented. Ways of complement are different in each used in the technique of projections methods.

1.3 Axonometric projection
1.3.1 A basic concept

Projection of figures with the projections of the rectangular coordinates axes connected with figures is considered in an axonometric projection. Fig. 1.6 shows the projection of point *A* on a plane π '. This plane is used for the plane of axonometric projections. The point is attributed to a coordinate system *Oxyz* previously. The arrow showed a direction of projection. The direction should not be parallel to coordinates axes. The projection (A'_2) of point A_2 (a rectangular projection of the point on plane *Oxy*) shown together with the projection of the point *A*. The straight lines x, *y*, *z* are axes in space, and the lines *x* ', *y* ', *z* ' are their projections on the plane π' (the axonometric axis). The point A'_2 is called the *secondary* (more) projection of *A*. $O'A'_xA'_2A$ ' is an axonometric projection of the coordinated polyline OA_xA_2A.

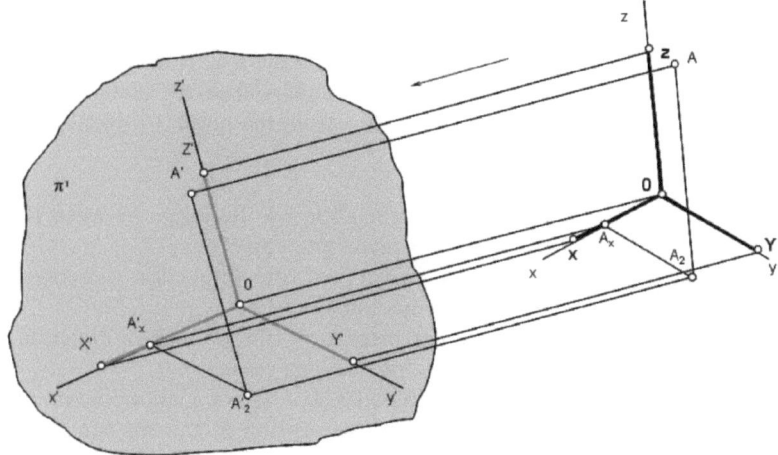

Fig.1.6

Take notice that the axonometric projections of parallel in space lines will be parallel. Measure off and put the segments $|OX| = |OY| = |OZ|$ on the axes x, y, z from the beginning of the coordinates . These segments are take as units of measurement and are called a *natural unit*. Segments $[O'X']$, $[O'Y']$, $[O'Z']$ are axonometric projections of natural units. In general, they are not equal. These segments would be accordingly axonometric units – the units of measurement along axonometric axes. Relations

$$\frac{O\prime X\prime}{OX}, \frac{O\prime Y\prime}{OY}, \frac{O\prime Z\prime}{OZ}$$ are called the *distortion factors* on axonometric axes. There are denoted as k_x, k_y, k_z accordingly with axes x, y, z.

1.3.2 The basis theorem of axonometric

A set of types of axonometric projections one can receive depending on a mutual position of a natural coordinates system and the plane of the axonometric projections, and a direction of the projection. They are differed from each other with the direction of axonometric axis and a quantity of distortion factor on these axes.

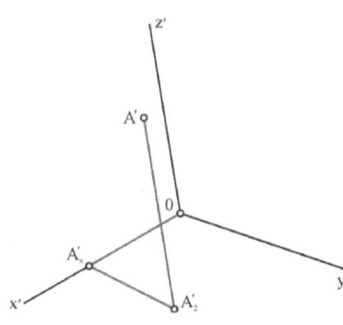

Consider the theorem which was proved by K.Polke: *three segments of arbitrary length lying in one plane and out from one point at arbitrary angles to each other are the parallel projections of three equal segments laid on the rectangular coordinate axes from the coordinate beginning.* This theorem is called a *fundamental theorem of parallel axonometric.* An axonometric projection of any figures one can construct with help of the projections of belonged to them points selected the distortion factors and the angles between the axes.

Fig.1.7

5

Consider the construction an axonometric projection of the point A (x_A, y_A, z_A). Draw the axis x', y', z' in any direction from the point O' and a polyline $O'A'_xA'_zA'$ (Fig. 1.7). Measure and put the segment $O'A'_x$ on the x'-axis from point O' (a length of the segment $|O'A'_x| = k_x x_A$), the segment $A'_x A'_2 -$ from the point A'_x parallel to the axis y' ($|A'_x A'_2| = k_y y_A$); the segment $A'_2 A' -$ from the point A'_2 parallel to the z'-axis ($|A'_2 A'| = k_z z_A$).

1.3.3 Types of axonometric

The axonometric are divided depending on the angle between the direct of projection and axonometric projections plane into two types:

1. *Rectangular* axonometric projection (the projection direction is located perpendicularly to the axonometric projections plane) .

2 . *Aslant-angular* axonometric projection (the projection direction is located not perpendicularly to the axonometric projections plane) .

If all three distortion factors are equal ($k_x = k_y = k_z$), axonometric projection is called an *isometric* projection. If only two distortion factor are equal to each other (e.g. , $k_x = k_y \neq k_z$ or, $k_x \neq k_y = k_z$), the projection is called *dimetric*, and if $k_x \neq k_y \neq k_z$, it is called *trimetric*.

STST 2.317-69 recommends to use five types of axonometric on the drawings used in all sectors of industry and civil engineering. They include: two rectangular (isometric and dimetric) and three aslant-angular (frontal and horizontal isometric and frontal dimetric) . These types of axonometric are called the *standard* axonometric.

1.3.4 Relationship between the distortion factors and the direction of projection

The angle between the projecting straights and the projection plane is denoted as φ°. The distortion factors and the angle that defines the projection direction are connected as follows:

$$k^2_x + k^2_y + k^2_z = 2 + ctg^2\ \varphi^\circ \qquad (1.1)$$

In a rectangular axonometric is: $k^2_x + k^2_y + k^2_z = 2$, $\qquad (1.2)$

because in formula (1.1.) $\varphi^\circ = 90°$ and $ctg 90° = 0$.

1.3.5 The standard axonometric

1.3.5.1 The standard rectangular axonometric

1) in the rectangular isometric each of distortion factors is equal to approximately 0.82 in accordance with the formula (1.2.) and the equality $k_x = k_y = k_z$. An equality $cos\alpha° = cos\beta° = cos\gamma°$ is received from a equality of distortion factors, i.e. the angles between the axes is equal to $360°:3 = 120°$ (Fig.1.8, a).

2) STST 2.317-69 recommends to take $k_x = k_z$, $k_y = 0,5\ k_x$ in dimetric. From formula (1.2) is find that $k_x = k_z \approx 0,94$; $k_y \approx 0,47$. The angles between the axes are: $(x' \wedge z') = 97°10'$, $(x' \wedge z') = 180° - 48°35' = 131°25'$ (Fig. 1.8, b).

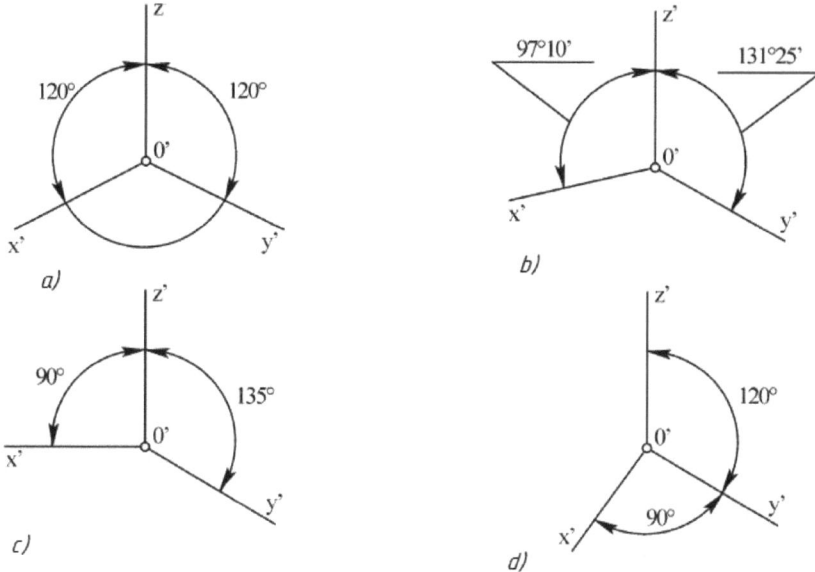

Fig.1.8

Pay attention to the fact that previously found distortion factors are very inconvenient to use, so the suitable coefficients to use instead of them it are advisable. For example, $k^*_x = k^*_y = k^*_z = 1$ is taken in the isometric, and $k^*_x = k^*_z = 1$, $k^*_y = 0.5$ – in the dimetric.

1.3.5.2 Standard aslant-angular axonometric

1) $k_x = k_z = k_z = 1$ is in the standard aslant-angular frontal isometric, $k_x = k_z = 1$ and $k_z=0.5$ is in the standard aslant-angular frontal dimetric, and a situation of axonometric axes for both cases are shown in fig. 1.8 c . The (x'^ y') one can take equal to 120 ° or 150 °.

2) $k_x = k_z = k_z =1$ is in the standard aslant-angular horizontal isometric, and the axes are located as shown in fig. 1.8 d. The (z '^ y') one can take equal to 135° or 150°, but the (x '^ y') must be equal to 90 °.

Questions and exercises

1. How is central projection formed?
2. Explain the concept of reversibility of a drawing.
3. What is an axonometric projection called?
4. What is the difference between the rectangular and aslant-angular axonometric projections?
5. What is an essence of the fundamental theorem of axonometric?
6. Name the types of axonometric.
7. What axonometric projections are standard?
8. What is the distortion factor?

Chapter 2. RECTANGULAR PROJECTION

The method of a rectangular projection on two mutual perpendicular planes is method frequently used for receiving reversible image. The planes are taken for a plane of projection. This method provides the accuracy and ease of measurement on the images of objects on plane, and it is the primary method for drawing of the technical drawings. The method was made at the end of the XVIII century by the French scientist G. Monge (1746-1818).

2.1 Projection of a point on two planes of projections

Take two mutual perpendicular planes. Dispose one of them vertically, and the second plane – horizontally. First plane is called *a frontal plane of projection* and is denoted as π_1, and the second plane is called *a horizontal plane of projection* and is denoted as π_2. The line of planes intersection is a *projection axis* and it is denoted as x (Fig. 2.1).

Construct a rectangular projection of the point A on these planes. For this draw from this point the perpendiculars to the planes of projections π_1 and π_2 and find their foundations: A_1 (frontal projection of point A) and A_2 (horizontal projection of point A). In order to go from the axonometric projection shown in the picture to the flat drawing, remove the object (point A) and leave alone its projections. However, connect these projections each other at first. Then a situation of the point in space will be able to determine. It is not difficult to understand that in order the projections to tie, the axis x to use it is advisable. So as the projections to tie with an axis x, it is necessary the perpendiculars to draw from the points A_1 and A_2 to the axis x. Take note of the fact that the perpendiculars are intersected in the x-axis at the one point (see Fig. 2.1).

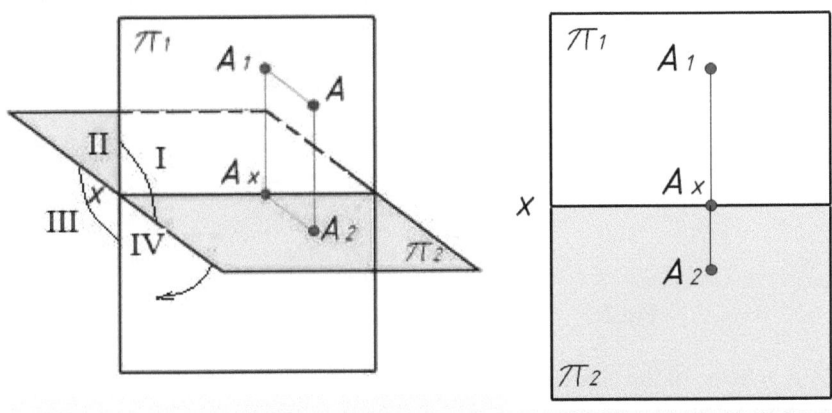

Fig.2.1   Fig.2.2

This point is called a *projection of point A on the x-axis* and is denoted as A_x. Receive a single plane (a *plane of the drawing*) turning around the x-axis the plane π_2

to coincidence it with the plane π_l (direction of rotation is shown in Fig. 2.1). The points A_1 and A_2 will be located on a line perpendicularly to the axis of projections. This perpendicular (straight line A_1A_2) is called a *line of tie*. The drawing received in consequence of coincidence of the planes is called the *complex drawing* or *Monges epure* (Fig. 2.2). The Monges epure will be used mainly further, these complex drawings will be called «drawing» or «epure», the border of projections planes will not be shown and projections planes will not be designated.

We can assert that the drawings constructed with Monges method are reversible. Indeed, find a situation of the point A in space in the fig. 2.2: turn the plane π_2 in its place, draw from the points A_1 and A_2 the perpendiculars to the planes π_1 and π_2. Point A will be locating in an intersection of these perpendiculars. Projection planes π_1 and π_2 divide the space into four parts (quarters). They are numbered as shown in Fig. 2.1.

2.2 Projection of a point on three planes of projections
Introduce third plane in the system of planes π_1 and π_2. This plane is denoted as π_3. Plane π_3 is located vertically and perpendicularly to the plane π_1, and to the plane π_2 (Fig. 2.3). The axis y and z are formed in the intersection the plane π_3 with planes π_1 and π_2. The point of intersection of projections of all three axes is denoted

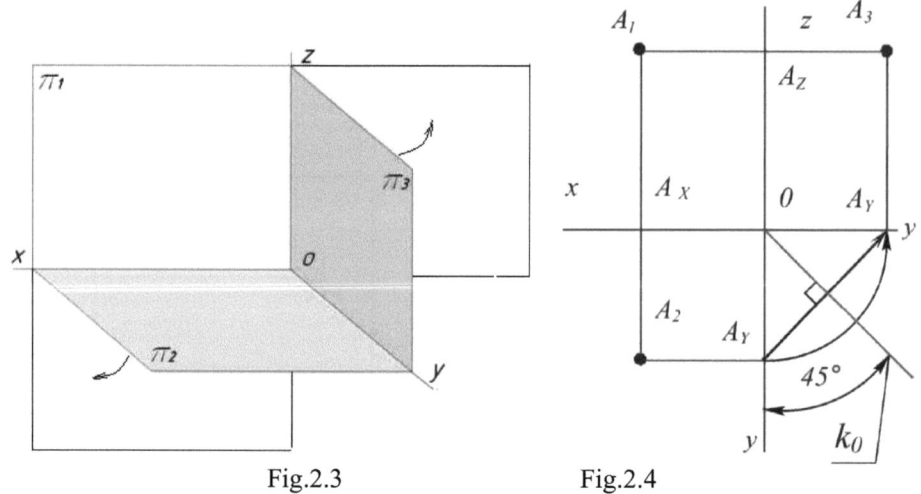

Fig.2.3   Fig.2.4

as letter O. A scheme of the coincidence the planes π_1, π_2, π_3 into one plane is shown in fig. 2.3. Two positions for the y-axis are given. A progress of finding a point A_3 using the frontal and horizontal projections is shown in fig. 2.4: one can use either a circular arc drawing from the point O, or bisector line of angle yOy – line k_0 (it is called a *constant of complex drawing*).

2.3 Projection of a line on epure

The property of parallel projection is known: "a straight line is projected in a straight line". Such conclusion one can make from this: to construct a projection a straight line, it is necessary to have the projections of two points, lying on its **or** the projections of one point on the line and the angles of inclination the projections to the axis. For example, if $A \in a$, $B \in a$, then $a_1 = (A_1B_1)$, $a_2 = (A_2B_2)$ (Fig. 2.5). On the fig. 2.6 is used that $A \in a \Rightarrow A_1 \in a_1$, $A_2 \in a_2$, and that the projections of the line parallel to the axis x.

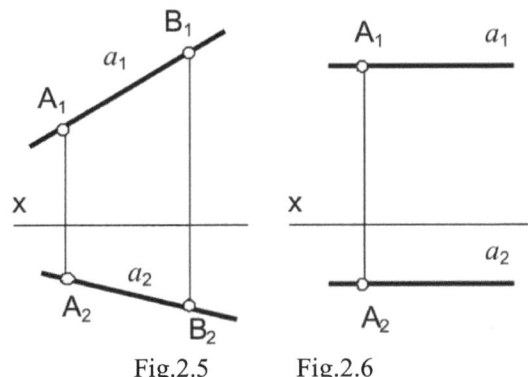

Fig.2.5   Fig.2.6

2.4 Situation of straights relatively of projections planes

2.4.1 The straights parallel to a planes of projections

The straights are divided depending on the situation of relatively of projection plane into frontal, horizontal and profile line.
1. The *frontal straight* – straight is parallel to a frontal projections plane. Its frontal projection is inclined to the axis of projections, and other projections will be parallel or perpendicularly to axes: $f \parallel \pi1 \Rightarrow f_2 \parallel x, f_3 \parallel z$ or $f_2 \perp y, f_3 \perp y$. If $[AB] \in f$, then $[A_1B_1] = |AB|$, and $(f_1, ^\wedge x) = \angle\alpha° = (f, ^\wedge\pi_2)$ (Fig. 2.7).

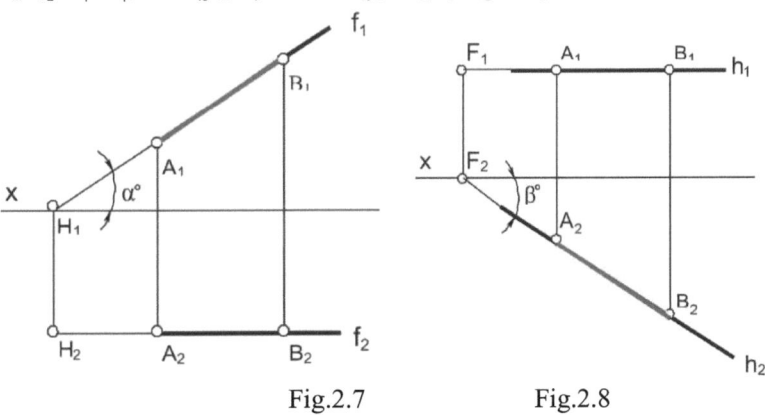

Fig.2.7   Fig.2.8

11

2 . The *horizontal straight line* – straight is parallel to a horizontal projections plane. Its horizontal projection is inclined to the axis of projections, and other projections are parallel or perpendicularly to axes : $h \parallel \pi_2 \Rightarrow h_1 \parallel x$, $h_3 \parallel y$ or $h_1 \perp z_1$, $h_3 \perp z$. Projection of segment lying on this line is equal to its life-size: $[A_2B_2] = |AB|$ (Fig. 2.8). On the epure the angle $\angle \beta^\circ = (h \wedge \pi_1)$ one can find without constructions.

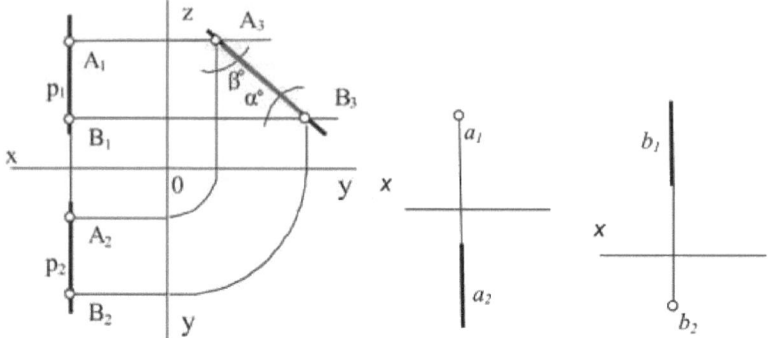

Fig.2.9   Fig.2.10   Fig.2.11

3. The *profile straight line* – straight is parallel to a profile projections plane. The profile projection of the line is inclined to the axis of projections, and other projections will be parallel or perpendicularly to axes : $p \parallel \pi_3 \Rightarrow p_1 \perp x$, $p_2 \perp x$ or $p_2 \parallel y$, $p_1 \parallel z$. If $[AB] \in p$, $[A_3B_3] = |AB|$, then $(p_3 \wedge y) = \angle \alpha^\circ =(p,\wedge\pi_2)$, $(p_3,\wedge z) = \angle\beta^\circ = (p,\wedge\pi_1)$ (Fig. 2.9).

2.4.2 The straight line perpendicularly to the plane of projection (projecting straight)

1. The *frontal-projecting straight* is the line perpendicularly to frontal projections plane. The frontal projection of this line is a point, and the rest will be located perpendicularly to the axes of the projections: $a \perp \pi_1 \Rightarrow a_2 \perp x$, $a_3 \perp y$ (Fig. 2.10) .

2. The *horizontal-projecting straight* is the line perpendicularly to horizontal projections plane. The horizontal projection of the line is a point and the frontal and profile projections are located perpendicularly axes: $b \perp \pi_2 \Rightarrow b_1 \perp x$, $b_3 \perp y$ (Fig. 2.11).

3. The *profile-projecting straight* is the line perpendicularly to the profile projections plane. The profile projection of this line will be a point, and the other will be located perpendicularly to the axes of the projections: $a \perp \pi_3 \Rightarrow a_2 \perp y$, $a_1 \perp z$. In the system π_1, π_2 horizontal and frontal projections of this line will be parallel to the axis x (the example is shown in Fig . 2.6).

2.5 Finding a life-size of a segment of line, the angles of inclination to a plane of projections

2.5.1 Finding the distance between two points.

In order to find a rule of determining the distance between two points with using of projections, consider their axonometric projections (Fig. 2.12,a). The distance between points A and B, i.e. a length of the segment AB, is a hypotenuse of

12

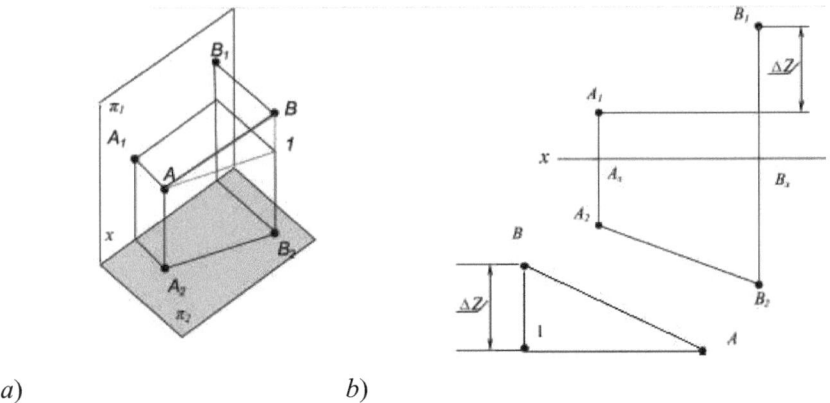

a)   b)

Fig.2.12

the triangle $AB1$, which is received when will be draw from point A the perpendicular to straight line BB_2. One leg ($A1$) of a right triangle $AB1$ is the distance between the horizontal projections of the points A and B, and the second leg ($B1$) equals the difference between the distances from these points to the plane π_2: [$B1$] = [B_2B] - [A_2A]). On the epure the segment $B1$ is equal to the difference in distance from the frontal projections of the points A and B to the axis x: [$B1$] = [B_1B_x] - [A_1A_x]) (Fig. 2.12,b).

2.5.2 Finding of the inclination angles of a straight line in general position to the projection plane

To find of the inclination angles of a straight line in general position to the projection plane, one can use this way: take on straight line the arbitrary segment, find a life-size with help of constructing a right triangle; then an inclination angle will equal to the angle between one leg and the hypotenuse of the triangle. Finding the life-size of the segment of line (the distance between two points) is shown in Fig. 2.12. However, from the constructed triangle $AB1$ is determined only inclination angle to a horizontal plane of projections ($\angle BA1$). To find an inclination angle to the frontal plane similar construction is performed, using the frontal projection of segment. Difference of segment endpoints (points A and B) to a frontal plane of projections ($\Delta y = [A_2A_x] - [B_2B_x]$) is take as a second leg. For example, it is necessary to find the inclination angles of straight a to the projections plane. Select on the straight the points A and B (it is necessary to forget that the projections of these points lie on single-valued projections of the line) first, construct the $\Delta A_2B_2B_0$ and define $\angle \alpha° = (a, {}^\wedge \pi_2)$. Select the segment A_2B_2 as one leg and take of the

13

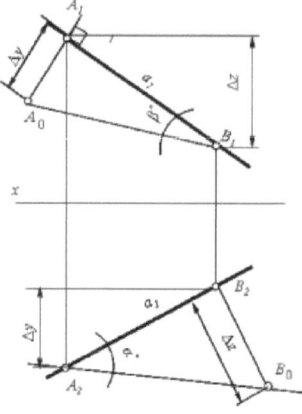

Fig.2.13

segment $[B_2B_0] = \Delta z$ as the second leg (Fig. 2.13). Select the segment A_1B_1 as one leg of a triangle, as the second – the segment $[A_1A_0]= \Delta y$, and construct $\Delta\ A_1B_1A_0$, from which define $\angle\beta° = (a, \wedge \pi_1)$. The equality $[A_2B_0] = [B_1A_0] =|AB|$ is the regularity.

2.6 The relative positions of two straight

Two straight lines in space can be located relative to the each other as follows:
a) to be parallel to each other;
b) to intersect;
c) to cross.

Based on the properties of parallel projection, one can say, that projection of these straight are located in the following way:
a) the projection *of parallel lines in general position* are parallel (Fig. 2.14);
b) the single-valued projections of *intersecting lines* is intersected, the intersection points lie on the one line of tie (fig. 2.15);

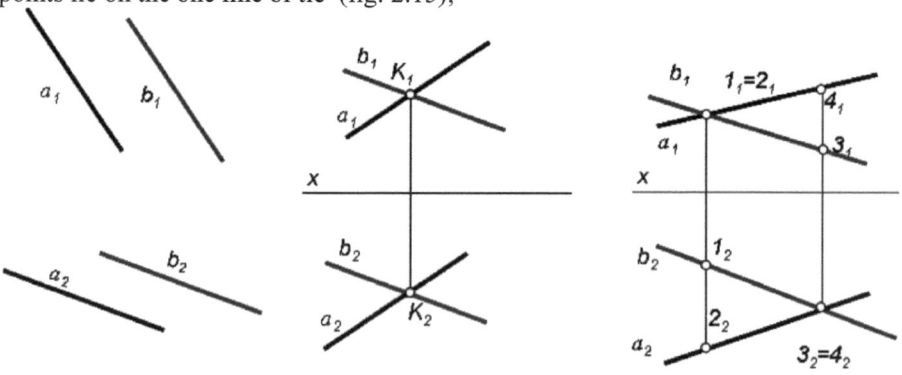

Fig.2.14   Fig.2.15   Fig.2.16

14

c) the projection of a *crossing lines* be able to intersect, but a point of intersection of the projections do not lie on the one line of tie (Fig.2.16). In this case, onto the epure competing (projecting each on other) points are appear.

Consider a special case of mutual intersection of the segments of general positions – the case of perpendicularity to each other. To find the distance between elements of space, it is necessary to draw a perpendicular. Therefore it is necessary to know how the projections of this perpendicular must be draw. Consider for this the theorem: *if one of the mutually perpendicularly straights is parallel to a plane of projections, then received on that plane projections will be mutually perpendicularly.*

Use the figure 2.17 for proof this theorem. Here the lines $a \times b$, $b \parallel \pi'$ and rectangular projection a' and b' on a plane π' are represented. Points A belongs to a, B belongs to b is taken on these lines except at a point K of their intersection. Prove that an angle $A'K'B'$ is equal to 90 degrees. We have that $K'K \perp \pi'$ and $b \parallel \pi'$ as a result of a rectangular projection. From the condition $a \perp b$ to get: $b \perp K'K$. Then the line b will be perpendicularly to two lines (AK and KK'), so it will be perpendicularly to the plane $A'AKK'$. As quadrangle $K'KBB'$ is a rectangle, then $b' \perp KK'$. We can say, that line b' is perpendicularly to the plane $A'AKK'$ use the deduction receive above. From this it follows that the line b', lying on the plane, is perpendicularly to the arbitrary

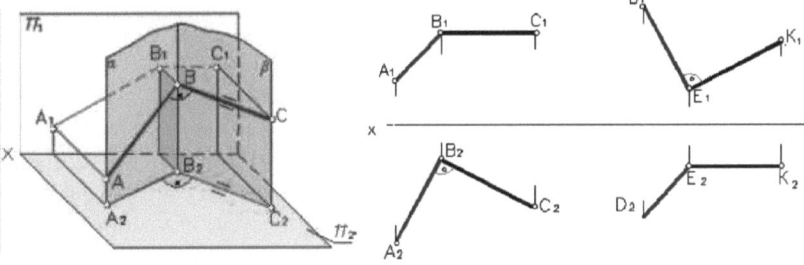

Fig.2.17   Fig.2.18

line passing through the point K' and lying in its, including to line a'. In other words, $a' \perp b'$, i.e. $\angle A'KB' = 90$ degrees.

Consider an example of application of the theorem. It is necessary to draw from point A the straight line a perpendicularly to straight line h (Fig. 2.18). As $h \parallel \pi_2$ (it one can determine from drawing on which $h_1 \parallel x$), straight lines a and h is projected on a plane π_2 at angle 90 degrees. Therefore $a_2 \perp h_2$ (straight line a_2 it is necessary to draw through the point A_2, as A belongs to a). Construct a frontal projection of line a used that it passes through the point A, and find another point of the straight a. Point K (the point of intersection of these lines) to take as this point correctly. Find the point $K_2 = a_2 \cap h_2$ at first, then draw a line of tie from the point K_2 to line h_1 and receive the point K_1, then connect it with the point $A_1 : (K_1A_1) = a_1$.

2.7 Construction of the traces of a straight line (line traces)

A point of intersection a straight line with the projections plane is called *trace*. On Fig. 2.19 a, it is shown the traces of the straight line a: H - *horizontal trace* of the straight line (the point of intersection of the line with a horizontal projections

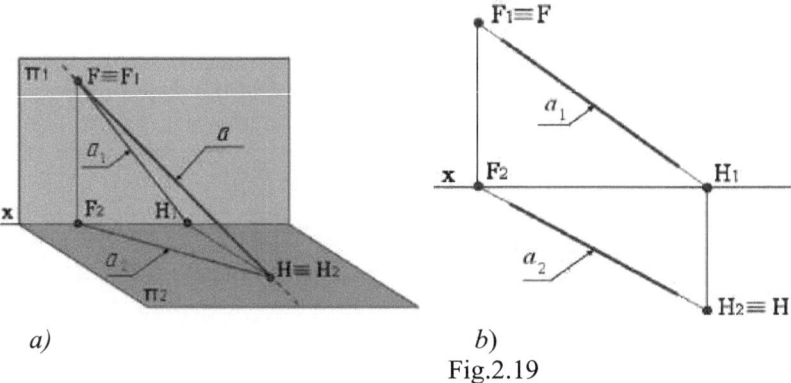

a)

b)

Fig.2.19

plane), *F - frontal trace* of a straight line (the point of intersection of the line with a frontal projections plane).

From this picture one can see: the horizontal projection of the horizontal trace of straight coincides with the trace, and its frontal projection lies on the *x*-axis; the frontal projection of the frontal trace coincides with the trace, and its horizontal projection lies on the *x*-axis. Therefore, to find the horizontal projection of the trace of the straight, it is necessary to find the point of intersection of the frontal projection with the *x*-axis : $H_1 = a_1 \cap x$; draw through this point the line of tie to its intersection with a horizontal projection of the line. Then the point $H_2 = H$ is determined (Fig. 2.19,b). To find projections of frontal trace of straight it is necessary to find the point of intersection of the horizontal projection with the *x*-axis : $F_2 = a_2 \cap x$; draw through this point the line of tie to its intersection with the frontal projection of straight. Then the point $F_1 = F$ is determined.

2.8 Ways of representing of the plane on epure

To determine a position of a plane in space it is necessary to have the following geometric element:
1. three points lying not on one straight line;
2. a straight and a point lying outside the straight;
3. two intersecting straight lines;
4. two parallel straight lines.

In accordance with this the plane can be give with the projections of the previously mentioned geometric elements on epure:
1) with projections of the three points lying not on one straight line;
2) with projections of straight line and point lying outside the straight;
3) with projections of two parallel lines;
4) with projections of two intersecting lines.

Each of shown ways on fig. 2.20 can be converted to another. For example, if through the points *B* and *C* showing on fig. 2.20 *a* a straight to draw, then is received

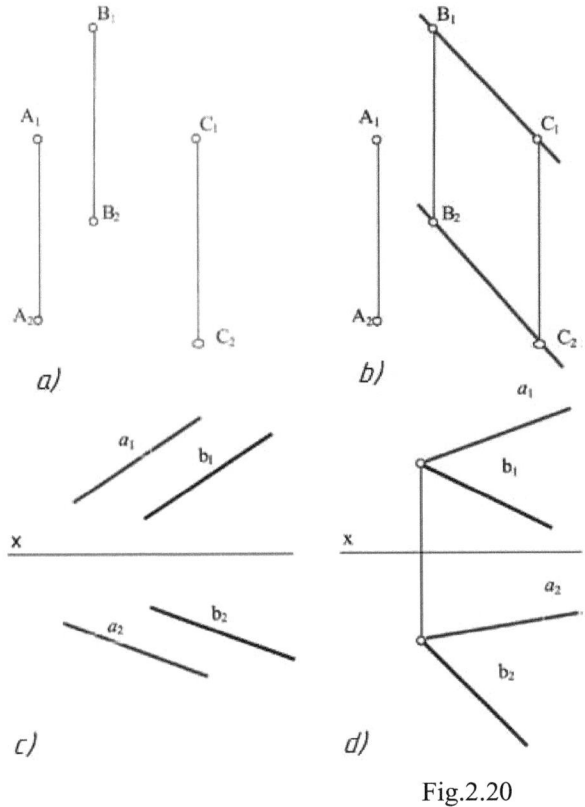

a)   b)

c)   d)

Fig.2.20

the way shown on Fig. 2.20,b. If three points is combined, then is received another way of plane representing – with the help of a flat figure (the triangle). The flat figure can be with a rectangle or other a polygon, circle and etc.

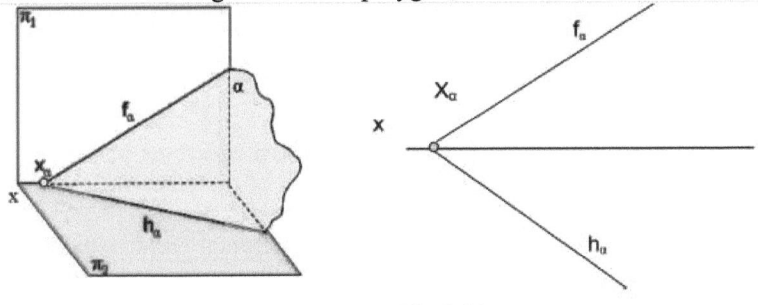

Fig.2.21   Fig.2.22

Consider a part of the plane α located in a first quarter of the space. This part of the plane is limited with the traces on the drawing, i.e. with the lines of intersection

17

of the projections planes and an arbitrary curve line (Fig. 2.21). Straight line, at which the plane intersects with the horizontal plane projections, is called the *horizontal trace* and is denoted as h_α. Straight line, at which the plane intersects with the frontal plane projections, is called the *frontal trace* and is denoted as f_α. In the drawing is clear see that the traces is intersected with each other, and that a point X_α is located on the x-axis. Point X_α is called *a point of converging of traces* (fig. 2.22).

2.9 The position of the plane relative to the projections planes
Planes are divided depending on their situation in space relative to the projections planes into two groups:
1. *The planes of the general position* - the planes is not parallel and not perpendicularly to projections planes.
2. *The planes of the private position* - the planes perpendicularly to one or two projections planes.

Such planes are called the *projecting planes*. One projection (trace) of projecting plane has a collective property, i.e. on this projection (traces) are projected

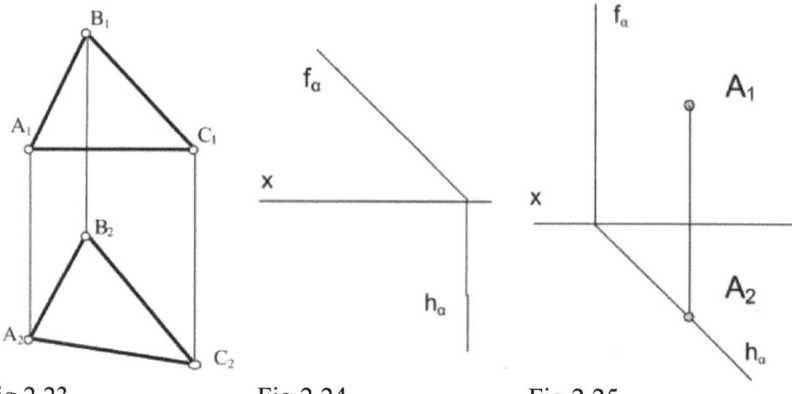

Fig.2.23   Fig.2.24   Fig.2.25

the points, lines, figures lying on the plane. The plane perpendicularly to two projections planes will be parallel to the third. One of the planes of general positions is shown in the previous paragraph; there it was given with traces. The plane of a triangle is shown on Fig. 2.23.

2.9.1 The planes are perpendicularly to one of the projections planes
The planes are called depending on of the situation of the plane relatively of projections plane so:
1. *Frontal-projecting plane* is a plane perpendicularly to a frontal projections plane. The frontal projection (frontal trace) this plane has a collective property, and the horizontal trace perpendicularly to the x-axis (Fig. 2.24).
2. *Horizontal-projecting plane* is a plane perpendicularly to a horizontal projections plane. The horizontal projection (horizontal trace) this plane has the collective property and the frontal trace perpendicularly to the x-axis (Fig. 2.25).
3. *Profile-projecting plane* is a plane perpendicularly to a profile projections plane.

The profile projection (profile trace) this plane has the collective property and on the complex drawing the horizontal and frontal traces are parallel to the x-axis.

2.9.2 The planes are parallel to the projections planes
Types of such planes are indicated below:
1. *Frontal plane* – a plane is parallel to the frontal projections plane. Each point of the plane is located on the same distance from the plane π_1, so the horizontal trace of it, as well as the horizontal projection, is parallel to the x-axis: $\alpha \parallel \pi_1 \Rightarrow h_\alpha \parallel x$. A segment of straight line or figure lying in this plane is projected on the plane π_1 in life-size: $\Delta A_1B_1C_1 = |\Delta ABC|$ (fig. 2.26).

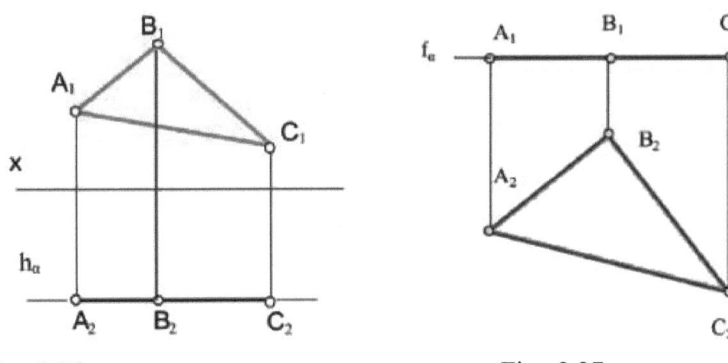

Fig. 2.26   Fig. 2.27

2. *Horizontal plane* - a plane is parallel to the horizontal projections plane. Each point of the plane located on the same distance from the plane π_2, so its frontal projection (along with her frontal trace) is projected in line parallel to the x-axis. A segment of straight line or figure lying in the plane is projected on the plane π_2 in life-size: for example, in fig. 2.27 $\alpha \parallel \pi_2 \Rightarrow f_\alpha \parallel x$, $\Delta A_2B_2C_2 = \Delta ABC$.
3. *Profile plane* – a plane is parallel to the profile projections plane. Horizontal and frontal projections (horizontal and frontal traces) this plane perpendicularly to the x: $\alpha \parallel \pi_3 \Rightarrow f_\alpha \perp x, h_\alpha \perp x$.

2.10 Projections of a point and a line located in the plane
2.10.1 The rules for constructing of the projections of a point and a line lying in the plane
These rules it are known from a course of elementary geometry:
1. If the point is located onto any straight line lying in the plane, then this point belongs to the plane (Fig. 2.28).
2. If two points of the straight line lie in the plane, then this straight line is located

Fig.2.28

in the plane. As a straight line is defined with two points, then the conditions of its situation onto plane are following:

2.1. If the straight line passes through two points on the plane.

2.2. If the straight line passes through one point of the plane and parallel to arbitrary straight line lying in this plane.

Based on these conditions a conclusion one can make: *for constructing of projection a line and the point lying in the plane, and for determining of the belonging of the line and the point with the plane, must be given the projections of the geometric elements defined the plane.*

2.10.2 Contour and frontour of plane

The *contour*[1] of plane is called straight lying on a plane and parallel to horizontal projections plane. As the line is parallel to the plane π_2, its frontal projection is parallel to the axis x. Therefore, the constructions of projections of contour it is necessary to start with its frontal projection. Its horizontal projection is found with the help of the projections of two points lying on the contour (fig. 2.29).

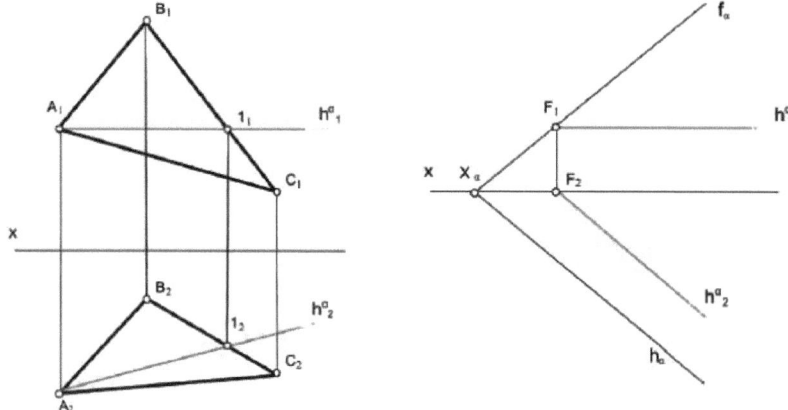

Fig.2.29   Fig.2.30

If the plane gives with traces, then the construction is much easier: only one point of contour (its frontal trace) is defined and the parallelism of contour and of horizontal trace is used (fig. 2.30): $h_1^\alpha \parallel x$, $h_2^\alpha \parallel h_\alpha$.

---

[1] term is take from a topographical drawing, where it is the closes line, received in consequence of section of ground with horizontal plane

20

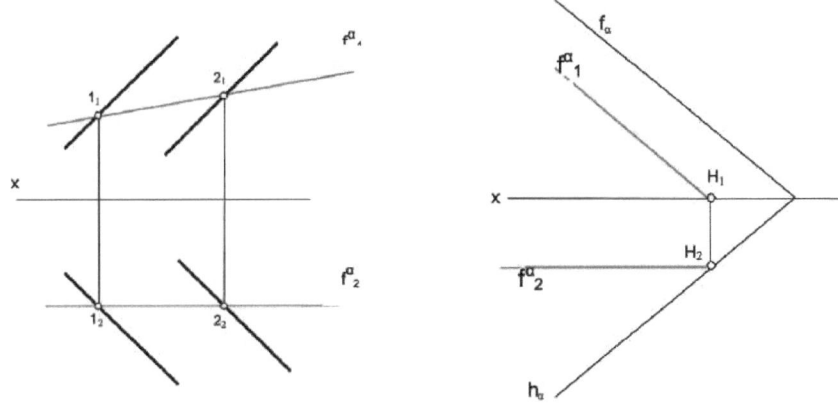

Fig.2.31   Fig.2.32

A *frontour*[2] of plane is called straight lying on a plane and parallel to frontal projections plane. As the line is parallel to the plane π_1, its horizontal projection is parallel to the x-axis. Therefore, the construction of projections it is necessary to start with its horizontal projection; its frontal projection is defined with projections of two points of the frontour (Fig. 2.31). If the traces of the plane are given, it is convenient to use the rule: a frontal projection of frontour is parallel to a frontal trace of plane: $f_2^a \| x, f_1^a \| f_a$ (Fig. 2.32).

2.11 The mutual position of two planes

Two planes in space can be is parallel to each other or is intersected.
1. The sign of a parallelism of two planes it is know from school: if two intersecting straight line of one plane is parallel to two intersecting straight

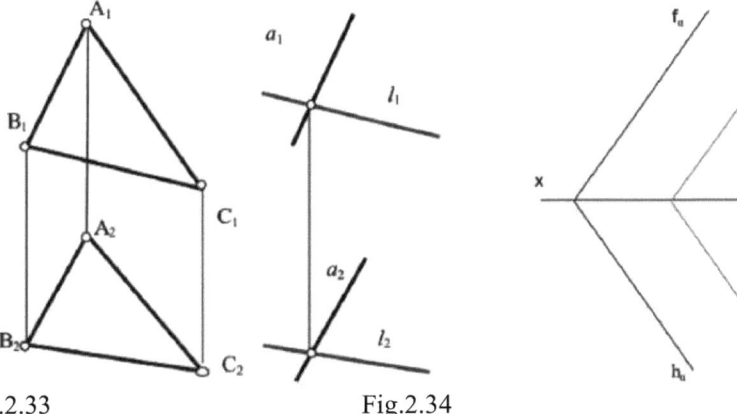

Fig.2.33   Fig.2.34

---

[2] The artificial term, taken consonant to *"contour"*

21

line of another plane, the plane is parallel to each other. True statement is inverse also, so one can write: $\alpha(a \cap b) \parallel \beta(c \cap d) \Leftrightarrow a \parallel c, b \parallel d$ (fig. 2.33). If the traces of a plane are taking as intersecting lines, then we receive: $\alpha \parallel \beta \Leftrightarrow h_\alpha \parallel h_\beta, f_\alpha \parallel f_\beta$ (Fig. 2.34).

2. A simplest case of construction on the epure a line of intersection two planes of the general position is case when the planes is represented with traces, and they is intersected in the field of drawing. In this case it's enough to find the points of intersection of traces: intersection line passes through these points: $(HF) = a$ (Fig. 2.35).

3. If

A) the plane is represented not only traces, or one - with traces, and the second with another way, **or**

B) the plane is represented with traces, but they do not is intersected in the field of drawing,

for constructing of the intersection line it is necessary to use one or two *intermediary planes* of private positions. It is necessary to pay special attention on

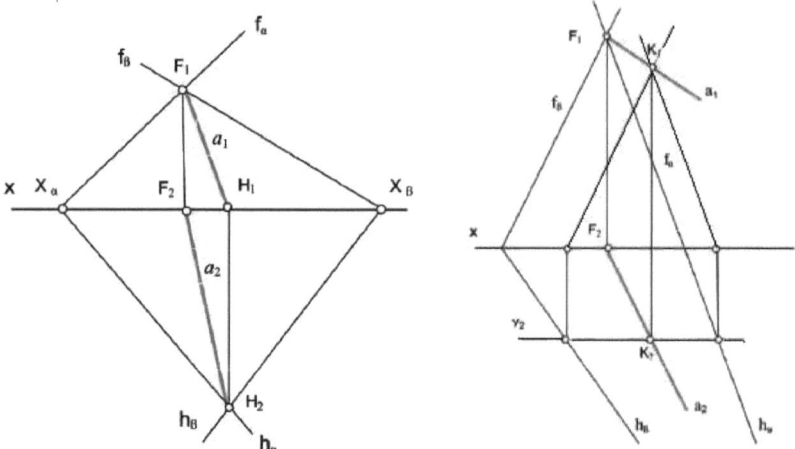

Fig.2.35   Fig.2.36

way of secant planes, and to understand its basic principle, because this way is one of the important ways of descriptive geometry. A main principle of the way consists of the following: the intermediary plane intersects the planes along straight lines; these straight lines is intersected with each other as lying in the same plane and give the general point for planes.

If you cannot find more one point, but with another way then it is necessary to draw another intermediary plane. For example, for constructing of a intersection line of the planes shown on Fig. 2.36, it's enough one intermediary plane to pass, because one point of a intersection line (point of intersection of the frontal traces) is determined easily. If the frontal plane (in the drawing it is plane γ) is take as the intermediary plane, then is determined point $K = (\alpha \cap \gamma) \cap (\beta \cap \gamma)$, and a line intersection is $a = (KF)$.

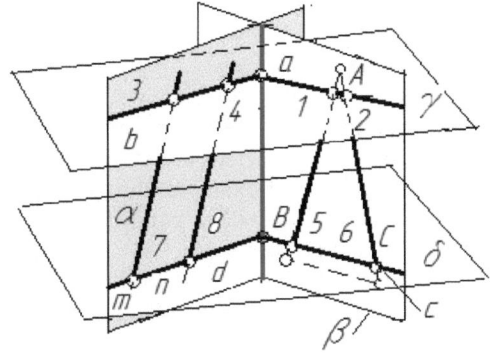

Fig.2.37

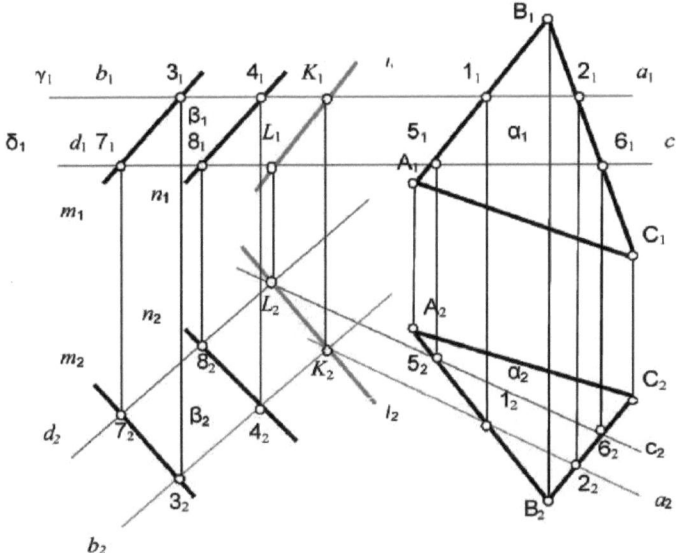

Fig.2.38

However, for constructing of the intersection line of the planes represented on fig. 2.37, it is not enough one intermediary plane. Draw the plane γ at first. This plane intersects the plane α ($\triangle ABC$) along the straight line $a = (12)$, plane β ($m \parallel n$) – along the straight $b = (34)$. In an intersection of these straights is defined the point $K = a \cap b$. Now draw another intermediary plane, define the straights $c = \delta \cap \alpha$ and $d = \delta \cap \beta$ and point L them intersecting ($L = c \cap d$). Then line $l = (KL)$ will be the intersection lines: $\alpha \cap \beta = l$. The solution this task on epure is shown on fig. 2.38.

2.12 Mutual position of a straight line and plane
Mutual position of a straight line and plane is possible in three cases:
a) a straight line lies in the plane;
b) a straight line is parallel to the plane;
c) a straight line is intersected with the plane.

A condition of belonging to plane a line it is known. In order the straight line was parallel to the plane, it is sufficient that it was parallel to any straight line lying in this plane. You know how the parallel lines are projected, therefore you can draw a projection line which parallel to the plane easily. For example, the projections of the line passing through the point K parallel to $\alpha(\triangle ABC)$, will be parallel to projections of arbitrary straight line of the plane (on fig.2.39 it is the line a).

To find a point of intersection of the line with a plane on the epure it is necessary to make the additional constructions. This task is solved easily only at an

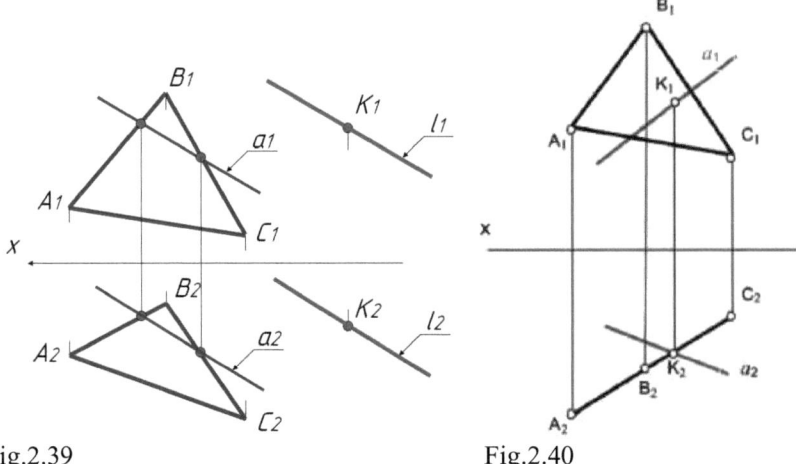

Fig.2.39   Fig.2.40

intersection of the straight line with the projecting plane. Indeed, one projection (trace) of projecting plane has a collective property; so the point of intersection is projected on this projection (on the trace). Therefore, one projection of the point will lay on place of intersection the projection of the straight line with projection (trace), which has the collective property; the second projection is finding with the help of tie line (fig. 2.40). On the picture the plane is adopted as transparent.

2.12.1 An intersection of the straight line with a plane of the general position
The simplicity of determination of intersection point the straight line with a plane of private provision has been shown above. But if the plane is a plane of the general position, then to find a point of intersection it is necessary to perform the following three steps (Fig. 2.41):

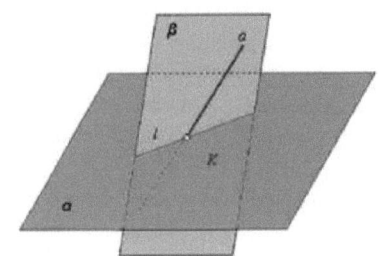

1) Pass an auxiliary plane through the given line. If its plane is projecting plane, then following construction is made easily.

2) Construct a line of intersection of the planes: given and auxiliary.

3) Determine the point of intersection of lines – given and constructed. The point will be the point of intersection of straight line and plane.

Fig.2.41

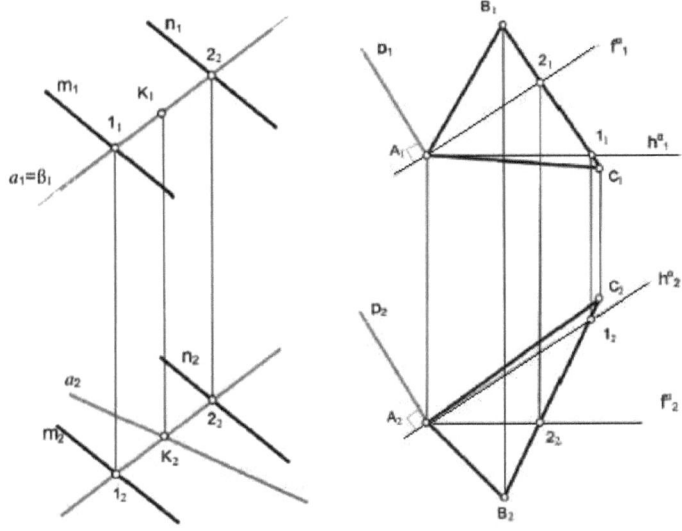

Fig.2.42   Fig.2.43

So, an algorithm of finding the point of mutual intersection of the line and the plane is written so: $a \in \beta \cap \alpha = b \cap a = a \cap \alpha$.

Solve the task in which it is necessary to find a point of intersection of the straight line and the plane is represented with two parallel straight lines (fig. 2.42). The frontal-projecting plane β is used as auxiliary intermediary plane; point K is found in intersection of the straight $(12) = \alpha \cap \beta$ with the given line a.

2.12.2 Perpendicularity a straight line and plane

Condition of perpendicularity a straight line and plane it is known from course of elementary geometry: the straight must be perpendicularly to two straight lines lying in the plane.

The contour and   frontour of plane to take convenient for fulfillment this condition, since the projection of right angles in space will be right angle. Then the

25

horizontal projection of the perpendicular will be perpendicularly to the horizontal projection of contour of plane, frontal projection of the perpendicular – to the frontal projection of frontour of plane: $l \perp \alpha \Rightarrow l_1 \cap f_1^\alpha$, $l_2 \perp h_2$ $^\alpha$ (Fig. 2 .43). Instead the contour and frontour of plane one can use the traces of planes: $l \perp \alpha \Rightarrow l_1 \perp f_\alpha$, $l_2 \perp h_\alpha$.

2.12.3 Perpendicularity of two planes

Two planes can be mutually perpendicularly, if at least one straight line lying in one of the planes, is perpendicularly to the second plane. Therefore, to construct the plane, perpendicularly to the plane, it is necessary to draw a perpendicular to the

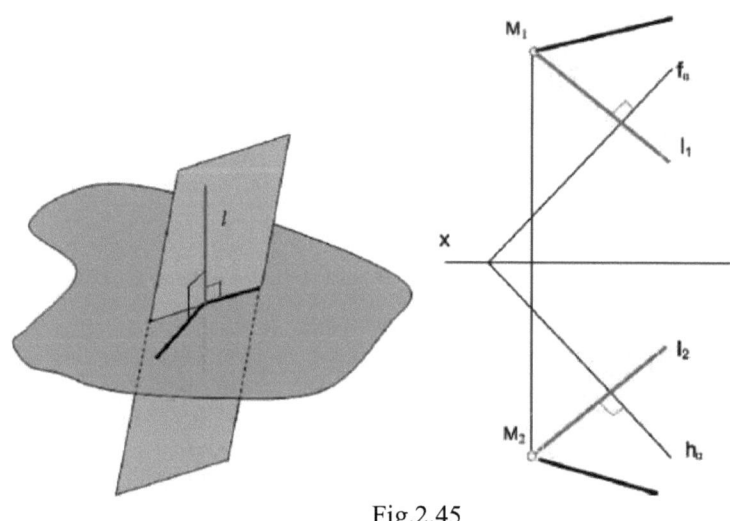

Fig.2.44   Fig.2.45

plane at first (Fig. 2.44). The projections of plane perpendicularly to the plane α are constructed on fig. 2.45.

2.13 Faceted surface
2.13.1 Rectangular projection of polyhedrons

Polyhedron is called the solid limited with flat polygons at which each side is one side of the other. These polygons are called *faces* of the polyhedron, their sides – *edges*, vertices – *vertices* of the polyhedron. A set of the facets is called a *faceted surface.*

Place of a polyhedron in the space is defined with different ways:
a) with the coordinates of the vertices,
b) with the base (if polyhedron is right and regular),
c) with one side and the number of sides (if polyhedron is correct).

Pay attention to a visibility of the edges of the polyhedron on the epure. The outlines of a projection of the polyhedron are always visible, and the visibility of the edges inside the outlines one can find with a way of competing points (Fig. 2.46).

2.13.2 An intersection of faceted surfaces with a plane

As a surface of the polyhedron consists out of flat elements, a line of its intersection with the plane is broken line. Therefore, in order to construct a line of intersection it is necessary to make following:

a) find the lines of intersection of a plane with the faces of surface (then the task will be reduced to a task of finding the line of mutual intersection of two planes); **or**

b) find the points of intersection of the plane with edges of the surface (then the task will be reduced to the task of finding the point of intersection of a straight line and a plane).

Example 1. Construct an intersection line of surface of the pyramid with the plane (Fig. 2.47). The give plane is projecting plane, therefore it is advisable to use the first

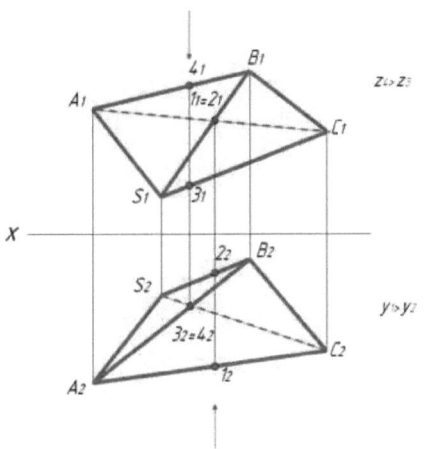

Fig.2.46

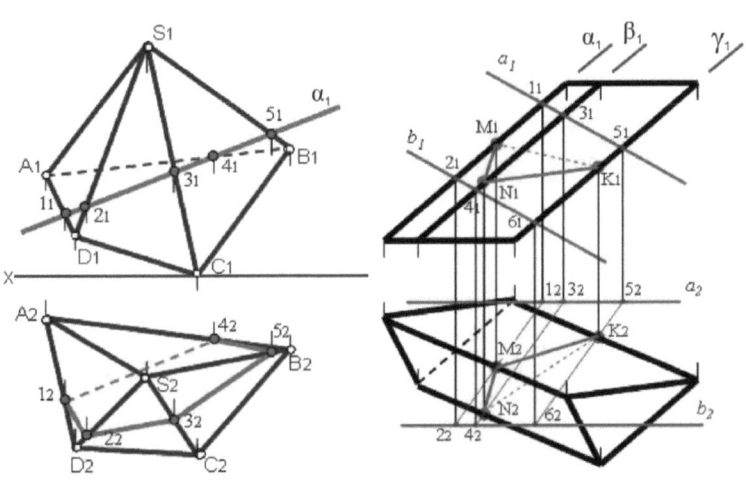

Fig. 2.47   Fig. 2.48

way: the points 1, 2, 3, 4, 5 are the points of intersection of the pyramid edges with the plane. On the epure the frontal projections of these points are found at first, and then their horizontal projections are constructed with help of the tie lines. The received points are connected according to the visibility.

Example 2. Construct an intersection line of the prism surface with the plane $\psi(a\|b)$; (Fig. 2.48). In this case, one can use the second way, and then the task of intersection of straight line with the plane is solved 3 times. The received points are connected according to the visibility.

2.13.3 An intersection of the straight line with a surface of a polyhedron

To find the points of intersection of the straight line with the surface three steps is performed:

1) construct a auxiliary secant plane through the straight line;
2) construct the line of intersection of the plane with the surface;
3) find the points of intersection of this line and the constructed intersection line. These points are the points of intersection of straight line and the surface.

Example: Find the points of intersection of the line a and the pyramid surface (fig. 2.49): $a \times \alpha = M, N$. Perform three describe above steps. Construct a auxiliary secant plane through the straight draw line a at first. It is advisable that it was projecting plane, and then its projection and projection of intersection line with the surface one can construct easily. In picture the auxiliary plane β is frontal-projecting plane, therefore $\beta_1 = a_1$. Now perform the second action: construct a line of mutual intersection the plane β and the surface. The intersection figure is polygon, therefore it is necessary to construct the vertices of figure. Frontal projection of vertices

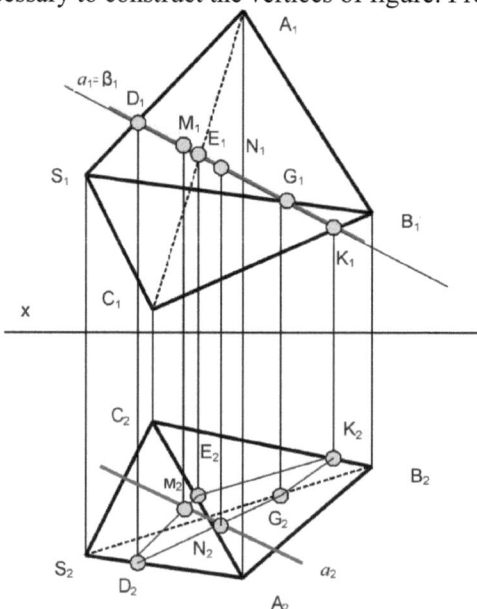

Fig.2.49

(points D_1, E_1, G_1, K_1) are defined easy, because the frontal projection of plane β has a collective property ($D_1 = [A_1S_1] \cap \beta_1$, $E_1 = [A_1C_1] \cap \beta_1$, $G_1 = [B_1S_1] \cap \beta_1$, $K_1 =$

28

$[C_1B_1] \cap \beta_1)$. To find the horizontal projection of points from the points D_1, E_1, G_1, K_1 draw the lines of tie to the segments $[A_2S_2]$, $[A_2C_2]$, $[B_2S_2]$, $[B_2C_2]$.

Then perform the third action: find the intersection points of the intersection line (of the polygon $DEKG$) with the given line. The horizontal projections of intersection points: $D_2E_2 \cap a_2 = M_2$, $D_2G_2 \cap a_2 = N_2$ is found first, and then used the lines of tie are constructed the frontal projections of points M and N. The visibility of the points of given elements on the drawing, one can find with way of competing points.

2.13.4 An intersection of faceted surfaces

Two faceted surfaces are intersected along one or two closed lines. To construct these lines, the points of intersection the edges of one surface with the faces of the second surface is found at first, and then the points of intersection the edges of the second surface with faces of first are determined. Combining the found points receive a broken line. Otherwise speaking, a solution of task of determining a line of intersection of the two faced surfaces is reduced to a solution of such tasks:

- find a intersection point of the line with the surface of the polyhedron;
- find a intersection line of the two planes (faces of polyhedron).

Two rules it is necessary to remember when connection of found points:

1. the points lying on one faces of the polyhedron one can connect only;
2. the line of intersection of two visible faces will be visible (if when intersection of the faces at least one will be invisible, then lying on this side the part of a line of intersection will be invisible).

Questions and Exercises

1. How the model of three mutually perpendicularly planes one can convert in the flat model?

2. Find the distance from point A (20,30,40) to the axes x, y.

3. Point A (50,30,40) is given. Construct the projections of point B located symmetrically to A relatively of the plane π_1, and of point C located symmetrically to A relatively of the plane π_2.

4. Construct three projections of profile straight line passing through point C (20,30,50), and at an angle of 60 ° with the plane π_1.

5. Find the distance between points A (10,30,10) and B (-20,5,10). Construct the projections the point C lying on the line AB and located at 30 mm from the point A.

6. The life-size of the segment AB of general position is 40 mm and its horizontal projection is located at an angle of 60 ° with the axis x. Construct the projections of the segment, if the coordinates of the point A are known: $x_A = 10$, $y_A = 20$, $z_A = 30$. One version of the task to solve enough.

7. Construct the tracks of straight line AB (see Exercise 5).

8. Determine the mutual arrangement of straight lines given on the epure (Fig. 1).

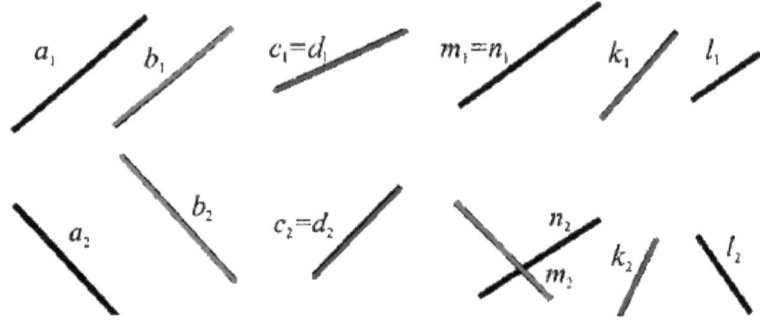

Fig.1

9. Construct from point A (-20,10,40) the perpendicular to the straight line passing through points B (10,30,20) and C (25,30,40).

10 . The base of an isosceles triangle ADE lies on the line BC (see Exercise 9), its life-size is equal to 30 mm. Construct the projections of triangle.

11. The frontal projection of the triangle ABC lying in the plane α ($a{\times}b$) is given. Construct its horizontal projection (Fig. 2).

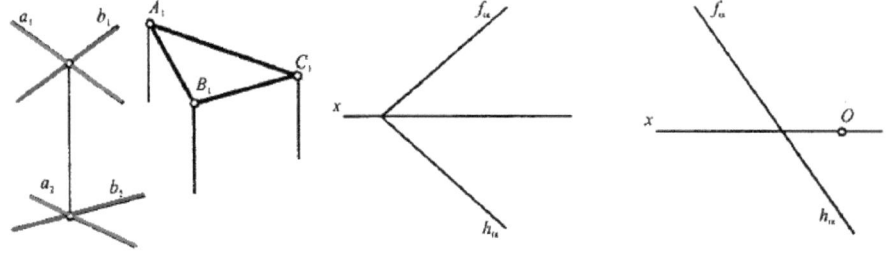

Fig.2   Fig.3   Fig.4

12. Construct the profile track of the plane α (Fig. 3).

13. Construct the tracks of a plane passing through the points A, B, C (see Exercise 9)

14. Construct the projections of bisector line of the angle between the tracks of plane α (Fig. 4).

15. Draw the frontour and contour of the plane passing through the points A, B, C (see Exercise 9)

16. Define that the points A (20,20,10), B (50,10, -10), C (-20,0,40), D (25,15,40) lie in a one plane or not.

17. Define the parallelism of the planes of arbitrary triangles ABC and LMN.

18. Construct the lines of mutual intersection of the planes which is given with the different ways (Fig. 5).

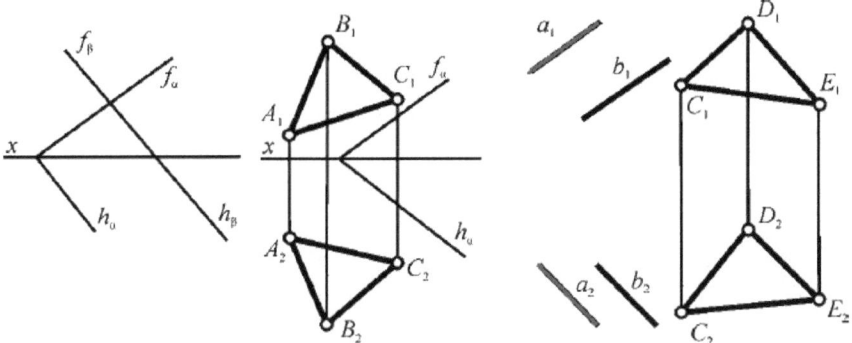

Fig.5

19. Draw from point A the straight line perpendicularly to the plane α (Fig. 6) .

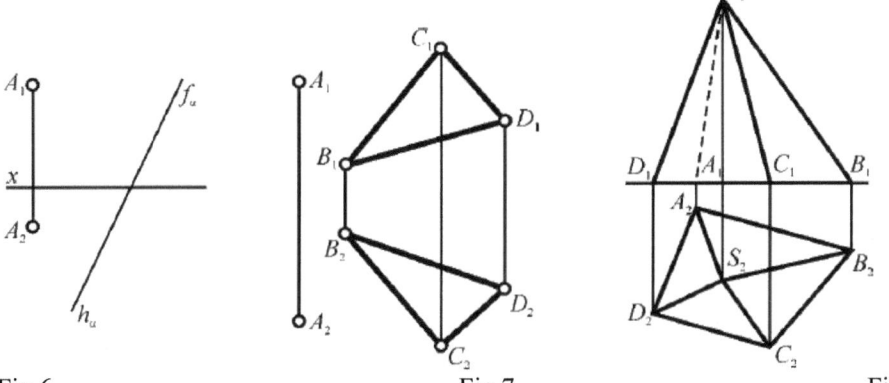

Fig.6   Fig.7   Fig.8

20. Find the distance from point A to the plane α ($\triangle BCD$) (Fig. 7).

21. Construct the projections of the intersection line of the pyramid surface with plane passing through the diagonal of its bases AC parallel to the edge BS, and the life-size of the section figure (Fig. 8).

22. In which cases, the given element is projected in life-size?

23. How is the distance between two points, two straight lines, straight line and plane determined?

Chapter 3. WAYS OF A TRANSFORMATION OF THE EPURE

3.1 Characteristic of ways of the transformation of the epure

Above it was noted that, if the straight lines and figures are in a private position concerning to the projection plane, the constructions for solving of task is relieved. Accordingly, it is necessary to know the ways of a transformation of the epure in order to pass from general position of geometric elements to private position them. *Transformation of the epure* is called a process of constructing with help of the given projections of object the new projections which it is convenient for solution of the task.

For transformation of the epure one can change the factors:

- a mutual situation of the geometry elements and projections planes; **or**
- a direction of projections.

The change of the situation of geometric elements in the given projection planes system can be accomplished using the following ways:

1) Way of *change of projection planes*: the receipt of private positions of the represented object with introduction of additional projections planes in a system of given projection planes;

2) Ways of *flat-parallel movement* and *rotation*: the bringing of object in a private positions with the changing its position in space.

The private positions of the object can be received with use of the additional parallel or central projection.

3.2 Way of the change of projections planes
3.2.1 Essence of way

An essence of the way of the change of projections planes is that the position of points, lines, flat figures, surfaces in space is remained unchanged, and the projections planes system is complemented with one or several planes consistently. The newly introduced projections planes must be perpendicularly to one of the projection planes of an old planes system.

Introduce the plane π_4 in the system of planes π_1, π_2. This plane must be perpendicularly to one of the planes π_1 or π_2, for example, π_2 (fig. 3.1).

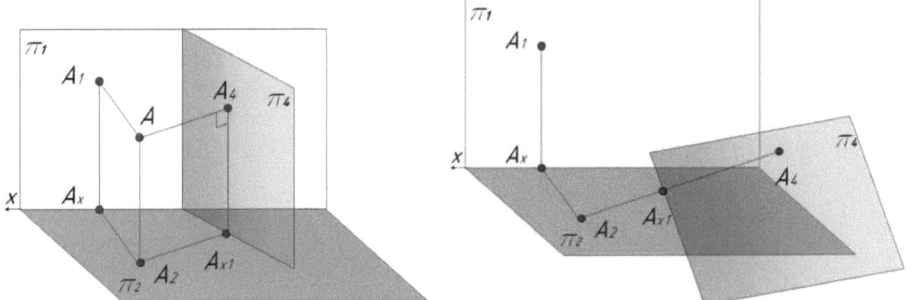

Fig .3.1   Fig .3.2

To find the projection of the point A in plane π_4, it is necessary to draw the perpendicular to this plane, and then to construct the point A_4. Fig. 3.2 shows the planes π_1, π_2, π_4 coincided with the same plane – the plane of the drawing; the resulting drawing is shown in fig. 3.3. To find the point A_4, it is necessary to draw the

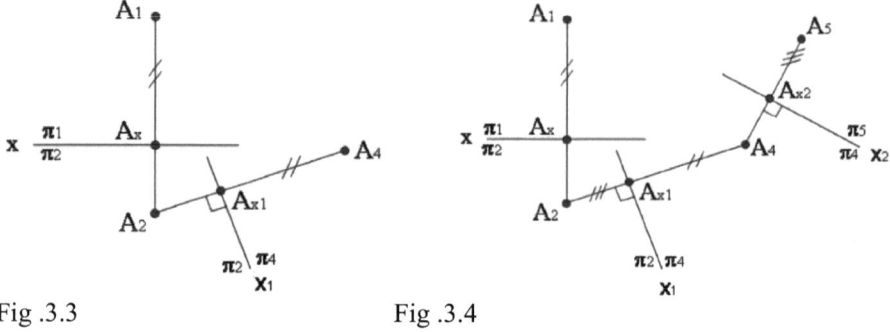

Fig .3.3   Fig .3.4

tie line perpendicularly to the axis x_1 from the point A_2, after that the segment $[A_{x1}A_4] = [A_xA_1]$ to measure and put from the point A_{x1}. That is clear from fig. 3.1, for the segments are equal: $[A_{x1}A_4] = [AA_2] = [AxA_1]$. Move from the received system of the planes π_2 ,π_4 to the system π_4, π_5. To find the projection of the point A on plane π_5 it is necessary to draw the line of tie perpendicularly to an axis x_2 through point A_4 , after that the segment $[A_5A_{x2}] = [A_2A_{x1}]$ measure off from the point A_{x2} (fig. 3.4). In this case is used this rule: the distance from the point to unchanged plane is not changed in the new system.

3.2.2 Examples of the use of the way
In the considered above example only one condition is done when introduction of additional projections planes (π_4 and π_5): the additional plane is perpendicularly to one of the projection planes. The additional requirements must fulfill for an achievement of the aim of projections transformation with the way of the projection planes change: the introduction planes must be parallel or perpendicularly to the given line or plane.
Example 1. To find the life-size of straight line segment of general position AB the additional plane π_4 must be parallel to this segment. Therefore on the complex drawing new projections axis x_1 mast be parallel to one of the projections, for example $[A_1B_1]$. The projections of points A and B on the plane π_4 (A_4 and B_4) are constructed in accordance with the said above rules. The length of the connecting them segment will be equal to life-size $[AB]$: $|A_4B_4| = |AB|$ (Fig. 3.5).

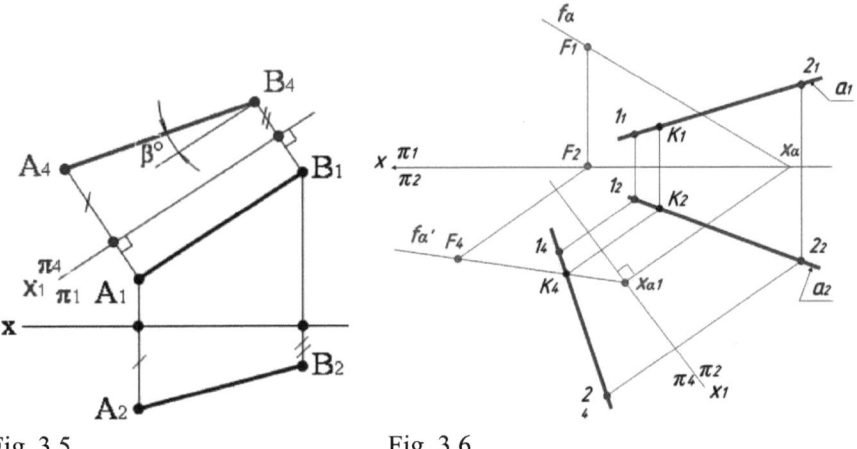

Fig .3.5   Fig .3.6

On the received drawing one can find the incline angle of the straight to the plane π_1: $((A_4B_4),^\wedge x_1) = ((AB)^\wedge \pi_1) = \beta°$.

If the x-axis to draw parallel to $[A_2B_2]$ then one can find the life-size of segment $[AB]$ and the incline angle to the plane π_2.

Example 2. If it is necessary to find the intersection point of the plane of general position and straight line a then instead of the three steps described above one can to consider the special position of the plane. This is a position of projecting plane in particular (Fig. 3.6). The trace of plane will have the collective property; therefore it is necessary to find the point K_4 at first. The points K_2 and K_1 are found with help of conditions of situation of the point on the line. A new projection of straight line a (a_4) is received with the help of auxiliary points 1 and 2. Their projections are found with the way shown above. To construct the track of plane (f_a') is used the point of converging of traces on the axes x_1 (X_{a1}) and the new projection (F_4) of point F lying on a frontal trace of the plane α.

Example 3. In the considered above examples the private positions of line and plane are received with the change of only one projections plane. In some cases the change of one plane will be not enough. For example, to find life-size of $\triangle ABC$ lying in a plane of the general position (fig. 3.7) the plane it is necessary to receive the projecting position at first, such the new projections plane to draw parallel to the plane of the triangle. In picture axis x_1 is perpendicularly to a horizontal projection of contour: $x_1 \perp h^\alpha{}_2$, therefore $\pi_4 \perp \alpha$, and the triangle is projected in the segment A_4C_4.

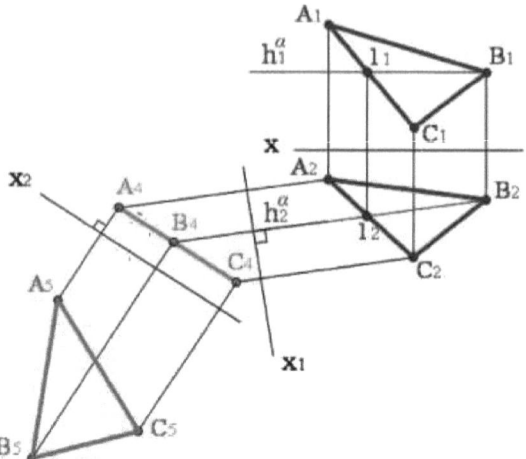

Fig.3.7

To plane π_5 was parallel to triangle ABC, the axis x_2 is drawn parallel to the segment A_4C_4. The new projection of the triangle is equal to its life-size: $\triangle A_5B_5C_5 = |\triangle ABC|$.

3.3 Way of rotation
3.3.1 Essences of way

When rotation around a fixed line (axis of rotation) each point of the arbitrary geometric element will be moved in the plane (*a plane of rotation*) perpendicular to the axis of rotation. The point moves along a circumference. Center of this circumference (*center of rotation*) is the point of intersection of the axis with the rotation plane; its radius is equal the distance from the rotating point to the center (*a radius of rotation*). If the points of the element lie on the rotation axis then these are immobile points. The rotation axis can be given or is selected. If the second case is used then the rotation axis must be located perpendicularly to one of the projection planes. Then the complexity of constructions will be decreased. Rotation of the surfaces, planes, lines lead to rotation of the points consisting of these elements. Therefore the essences of way need to consider on example of rotation a point.

1.   Let's the point A is rotated around the axis i which is located perpendicularly to the plane π_1 (Fig.3.8). The rotation plane α is parallel to the plane π_1 ($\alpha \perp i$, $i \perp \pi_1 \Rightarrow \alpha \parallel \pi_1$), therefore the trajectory of movement of the point (circumference) is projected on the plane π_1 as the circumference with the radius $R = | AB | = | A_1O_1 |$ (O - center of rotation) and center is at point O_1, on the plane π_2 it is projected as the segment of length $2R$ lying on the frontal trace of plane α. If the rotation axis is perpendicular to the plane π_2 then on this plane the circumference is projected as the circumference, and on the plane π_1 – as the segment is parallel to an axis x, therefore $\alpha \perp i$, $i \perp \pi_2 \Rightarrow \alpha \parallel \pi_2$.

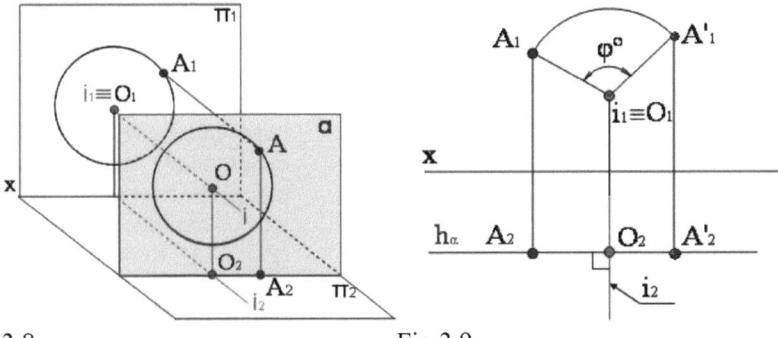

Fig.3.8   Fig.3.9

Fig . 3.9 shows the turning of the point *A* on the angle φ° around the axis *i* perpendicularly to the plane π₁. The arc of circumference is passed through point *A*. Its center is located ih the point O_1, central angle is equal to the angle φ°, radius $R = |A_1O_1|$. The point A'_1 is found at first, and the point A'_2 is found using the tie line. This line is passed through the point A'_1 to intersection with the line passing through the point A_2 perpendicularly to the line i_2 ($A_2 \in h_\alpha \perp i_2$).

2. Let's the point *A* is rotated around the axis *i* which is located parallel to the plane π₂ (Fig. 3.10). The rotation plane α is located perpendicularly to the plane π₂ (α $\perp i$, $i \parallel \pi_2 \Rightarrow \alpha \perp \pi_2$), therefore the trajectory of movement of the point

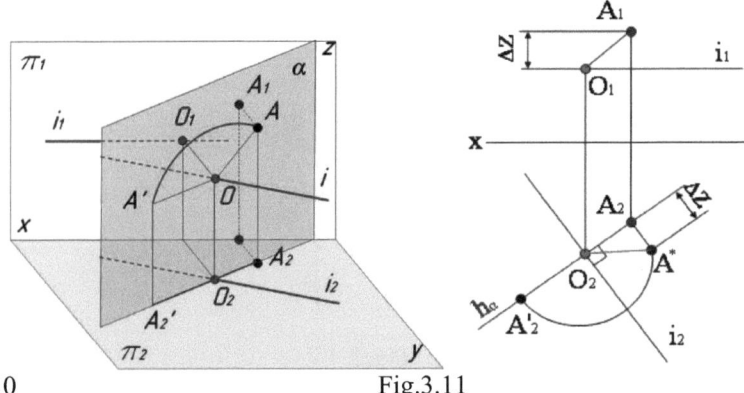

Fig.3.10   Fig.3.11

(circumference) is projected on the plane π₂ as the segment. This segment lies on the horizontal trace of plane α - h_α , and $h_\alpha \perp i_2$, because $i \perp \alpha$. The frontal projection of the circumference is an ellipse.

If the rotation radius comes to the position parallel to the plane π₂ then it is projected on this plane in life-size. To find this value, one can use the way of the

37

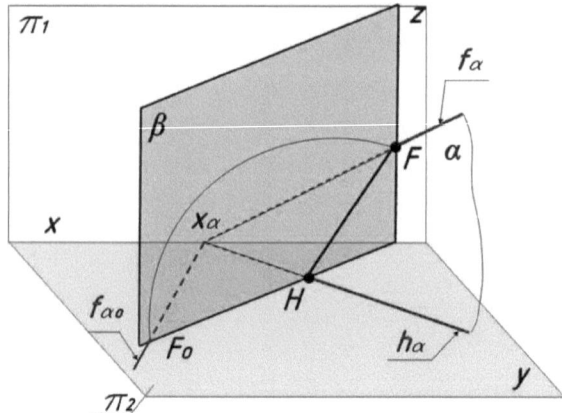

Fig.3.12

rectangular triangle. Take the segment O_2A_2 for one leg, segment $A*A_2 = \Delta z$ – for the second leg, then the hypotenuse O_2A* of the rectangular triangle will be equal to the radius, which it was necessary to find (Fig.3.11). In order to find the new position of the point A_2 (point A'_2), it is necessary to measure and put the segment $|O_2A_2| = |O_2A*|$ from the point O_2 on line h_α.

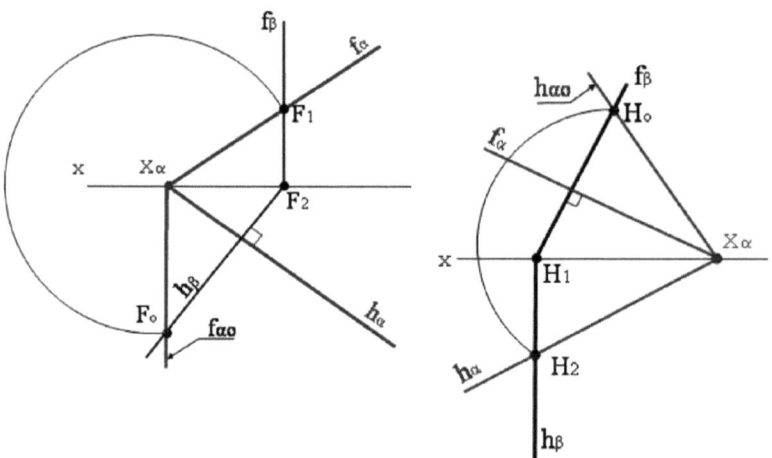

Fig.3.13   Fig. 3.14

In this example is considered the case when the point is rotated around the line i lying in horizontal plane. The way of rotation around an axis is parallel to one of the projections planes is used for finding a life-size of flat figures with rotation around a contour or frontour mainly. In this case the second projection of the trajectory of movement (ellipse) is not necessary to find.

If the plane is rotated around its trace to coincidence with the projections plane, in which this trace is located, then the figures lying in the plane are projected in the life-size on the plane. This construction is similar by its content to the rotation of a flat figure around the contour or frontour. In this case one trace of plane (in figures 3.12 , 3.13 it is horizontal trace) does not change its position. To find the coinciding position of the second trace it is necessary to use the point of converging of traces and the new projection of the arbitrary point (on the pictures - it's point F lying on the frontal trace). The sequence of constructions can be understood from fig. 3.14, in which the axis of rotation is located on the frontal trace of the plane.

3.3.2 Examples of use of the way

Example 1. Determine the life-size of the segment AB using the way of the rotation (Fig. 3.15). If the axis of rotation is not given, then his one can select. The easiest way is way when the axis is located perpendicularly to the plane of projection and is passed through outermost points of the segment. In the picture the rotation axis $i \in B$ and $i \perp \pi_2$, then the point B is remained stationary, and the projections of the point A will move as follows: A_2 – along circumference with center in the point B_2

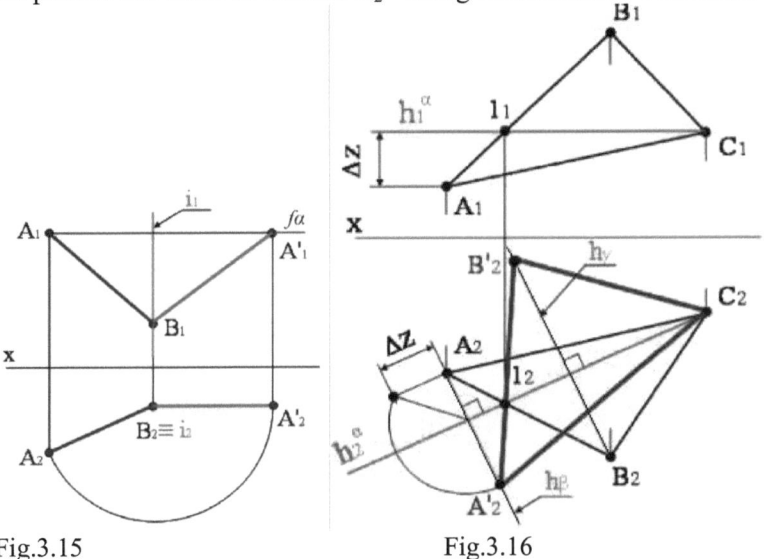

Fig.3.15   Fig.3.16

and with radius $R = |A_2B_2|$, point A_1 – along straight $f_\alpha \parallel x$. The segment must is projected on the plane π_2 in life-size, therefore the point A_2 it is necessary to rotate until the segment A_2B_2 does not come to the position parallel to the axis x. Then the new projection of [AB] will be [A'_1B_1] $= |AB|$.

Example 2 . Determine the life-size of $\triangle ABC$ with way of the rotation (Fig.3.16). If the flat figure is rotated around the contour and comes in the position parallel to the plane π_2, then it is obvious, that his horizontal projection will be equal to the life-size of the figure. Find the horizontal projection of the triangle. The points lying on the rotation axis are stationary points. Among them two points are necessary: the point C (one of the vertices of the triangle) and point 1. Above was proved, that points A_2 and B_2 are moved along lines h_β and h_γ, when is located

39

perpendicularly of the line $h_{a2} = i_2$. Now, finding the new projection of the point A (point A'_2) one can find the point B'_2 with help of the line passing through this point and the point I_2. In the result is received $\Delta A'_2B'_2C'_2 = \Delta ABC$.

Questions and Exercises

1. What is the difference between the way of rotation and the way of change the projection planes?
2. What is an essence of way of the flat-parallel movement?
3. How one can find the place of the rotation plane, the rotation radius, the rotation center when rotation around the contour or frontour?
4. How will be moved the projections of point when rotating around the axis perpendicularly to the horizontal plane of projections?
5. How many times is it necessary to make the flat- parallel movement in order that the straight line of general position come in horizontal-projecting position?
6. Find the distance between points A (20,15,30) and B (30 , -20,0) with way of change the horizontal plane of projection.
7. Find the distance between the straight lines a and b with way of change the projection planes (Fig. 1).
8. Find the angle between the planes α and β (Fig. 2).
9. Draw the plane parallel to the plane of ΔABC and located from it on the distance of 30 mm (Fig. 3).

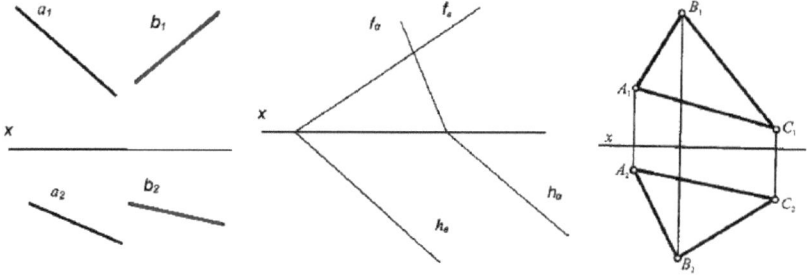

Fig.1   Fig.2   Fig.3

10 . Determine the distance between point A and the plane α (Fig. 4).

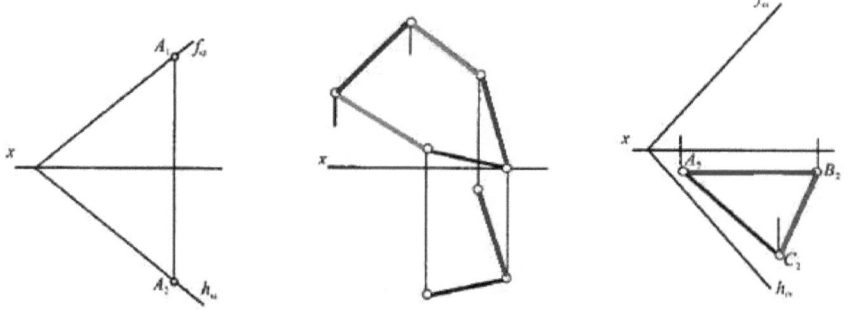

Fig.4   Fig.5   Fig.6

40

11. Solve the tasks 6, 7 partially change the conditions and using the way of rotation.

12. The frontal projection of the polygon and the horizontal projection of its two sides are given (Fig. 5). Finish the construction of its horizontal projection and find the life-size of polygon with way of rotation around the contour.

13. $\triangle ABC$ lying on the plane α is given (Fig. 6). Find the life-size of triangle with way of rotation around one of the tracks of plane.

14. Name three ways of finding of the life-size of the segment of straight line.

Chapter 4. CURVE LINES AND SURFACES

4.1 Projection of a curve line

In the base of the elementary concepts of descriptive geometry a curve line can be defined as a trajectory of moving of a point. The curve can be received as a result of mutual intersection of surfaces or intersection surfaces with a plane.

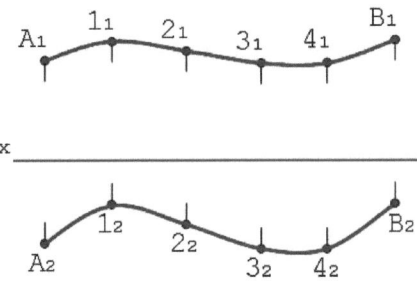

Fig. 4.1

To construct the epure of the curve line is necessary to construct the rectangular projections of points lying on it, then the projections of these points to connect with observance of the order of the points on the line (Fig. 4.1). Lines are divided in two groups: the *transcendental* (if they is defined with the transcendental equations) and *algebraic* (if they is defined with the algebraic equations). Lines can be space lines (if all points do not lie in one plane) or flat lines (when all points lie in one plane).

4.2 The cylindrical screw line

The screw lines are curves most used in the technique among of space. The screw line can be defined as a trajectory of the movement of the point

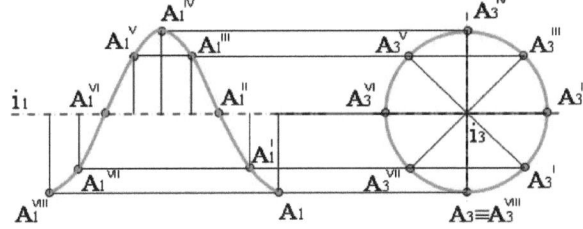

Fig. 4.2

around an axis in accordance with given regularity. For example, if the point is rotated around the axis and at the same time is moving along this axis, it describes a cylindrical screw line. If the speed of rotation of point is constant, then the cylindrical screw line is called *helix*. Fig. 4.2 shows the projections of one screw of the helix. The segment $A_1 A_1^{VIII}$ is called the *step of cylindrical screw line*.

4.3 The ways of forming of surfaces

Surfaces are the most common and infinitely diverse group of geometric figures. They include except a plane the complex forms of curved surfaces which don`t have exact mathematical description. The surfaces are not have unmatched among other geometric figures owing to diversity of forms and properties, their importance in the formation of different geometric figures, role in science, engineering, architecture, fine arts. In descriptive geometry the surface is considered as a collection of all successive positions of some line moving in space. The movable line is called a *generator*, and the lines defining the laws of movement are *guides*. The generator slides along guides.

Described way of forming the surface is called *kinematic*. In the kinematic way of forming of the surface great importance have the concept *"determinant"*, which is understood as a necessary and sufficient collection of geometric figures and ties between them, which the surface is defined uniquely.

The included in the determinant conditions are following:

1. List of geometric figures involved in the formation of a surface (point, line, etc.).
2. Algorithmic part which indicates on a relationship between these figures.

Thus, the determinant of the surface consists of two parts: the collection of geometric figures (first part) and additional information about the character of changing of the generator form and the law of its movement (second part). In order to distinguish the first (geometric) part of the determinant and the second (algorithmic) part one can do so: the first part to conclude in the round brackets, and the second – in square brackets. Then the determinant of the surface will have in the general case the following structural form : $\alpha\{(G), [A]\}$, where (G) is geometrical part, [A] - algorithmic part. To the determinant was refers to a specific type of surfaces it is necessary to enclose specific content into every part of the determinant.

Another way of forming of surfaces is way by which they are defined with a set of points or lines belonging to them. The points or lines are selected so that they give a chance to determine the surface form and to solve the various tasks. The ordered set of points or lines belonging to the surface is called its *framework*. The preliminary examination of orthogonal projections of the surfaces shows that some surfaces cannot be defined with their projections. This occurs because the surface can be the unclosed surface.

In order to give the surface on drawing, it is sufficient to indicate the projections not all set of points or lines belonging of the surface, and only some of them, with help which one can put the unambiguous correspondence between the projection and the object of projection. An outline of the surface on the projection plane one can draw also. Fig. 4.3 shows the surface α and the line l' when it its parallel projection on the plane π'. The curve l' is outline of surface. As it was shown above, the projecting straights are drawn tangency to the surface.

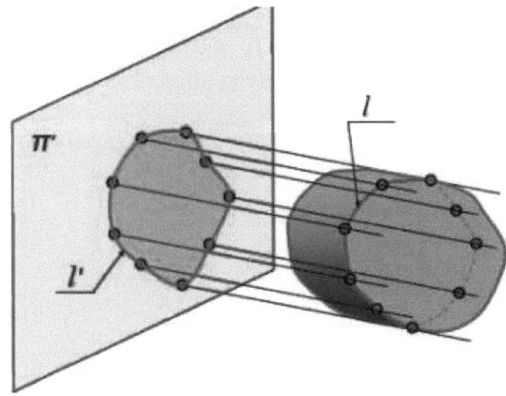

Fig . 4.3

The surfaces can be the linear and non-linear surfaces depending on type of generator. The generator of the first type is a straight line, and the second is a curve. Linear surfaces are divided into developable and non-developable. The technical surfaces are divided by a law of movement of a generating line and of forming surface into the groups:

1) surfaces of rotation;
2) screw surfaces;
3) surfaces with a plane of parallelism;
4) surfaces of parallel transporting.

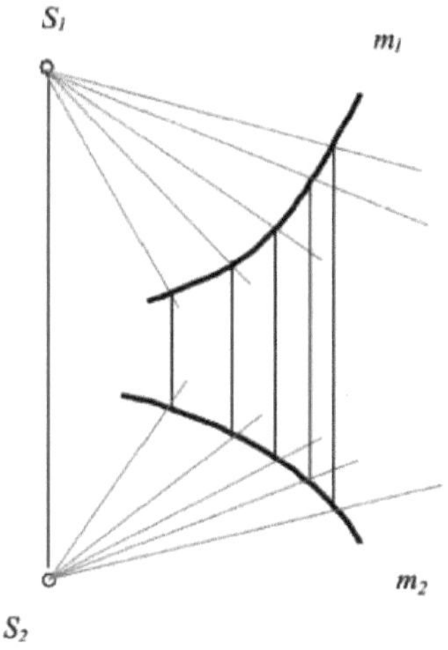

The significant class of surfaces is formed with the movement a circumference of constant or variable radius. They are called *cyclic* surfaces.

If a formation of a surface not corresponds to any particular geometric regularity, the surface is called *graphic surface*. The earth's surface is the graphic surface and is called a *topographic surface*. Topographic surface is imaged using collection of lines and points lying on the surface.

So, the world of surfaces is diverse and unlimited. Consider the image on the drawing the surfaces widely used in the technique.

4.4 Linear surfaces

As mentioned above generator of linear surfaces is straight lines. Their two types are known from the school:

a) conical surface. Geometric part of the determinant is the curve m, algorithmic part is the condition of passing of the generators through the point S when not lie on this line. In order to show the surface on the drawing the projections of m and S (the vertices of the conical surface) to construct it is enough. The projections of the several surface generators to show for clarity of image worthwhile (Fig. 4.4);

Fig. 4.4

45

b)   cylindrical surface. Geometric part of the determinant of surface are the curve *m* and direction of generators (straight line *l*), the algorithmic part is the condition of the parallelism of the generators to given direction (Fig. 4.5). This surface belongs to the surfaces of parallel transporting.

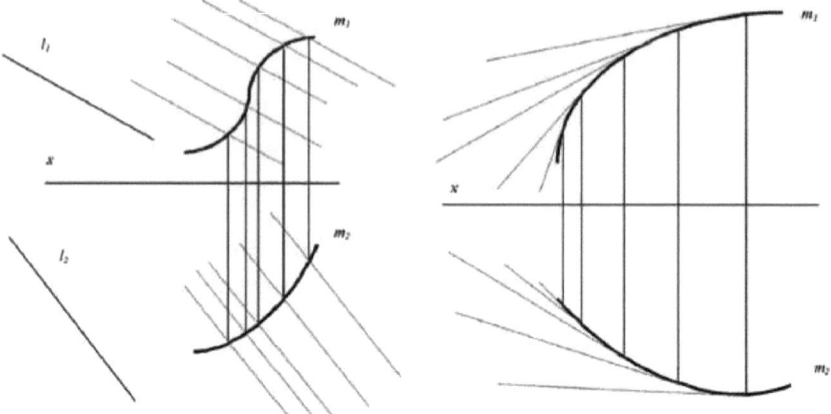

Fig.4.5   Fig.4.6

Conical and cylindrical surfaces are developable surfaces. There are only three types of developable surfaces. The third is the surface with an edge of return (torso). The generators of this surface pass tangency to the edge of return (this is an algorithmic part of the determinant) (Fig. 4.6).

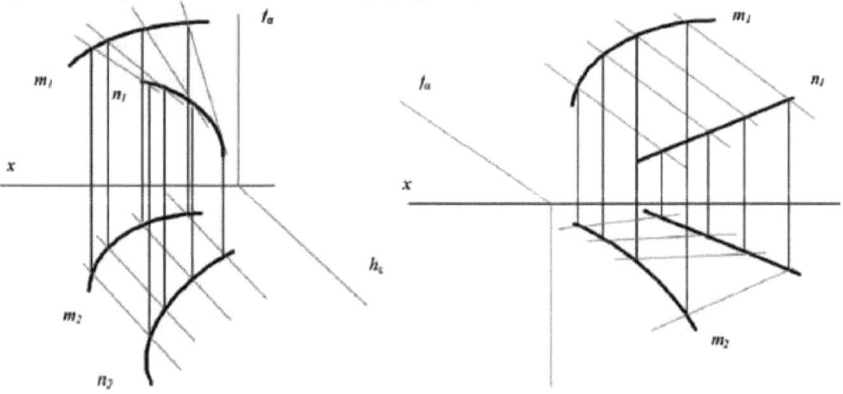

Fig.4.7   Fig.4.8

Linear surfaces are formed in consequence of movement of the straight line along the guide lines, therefore only one straightness generator passes through an arbitrary point of guide line.

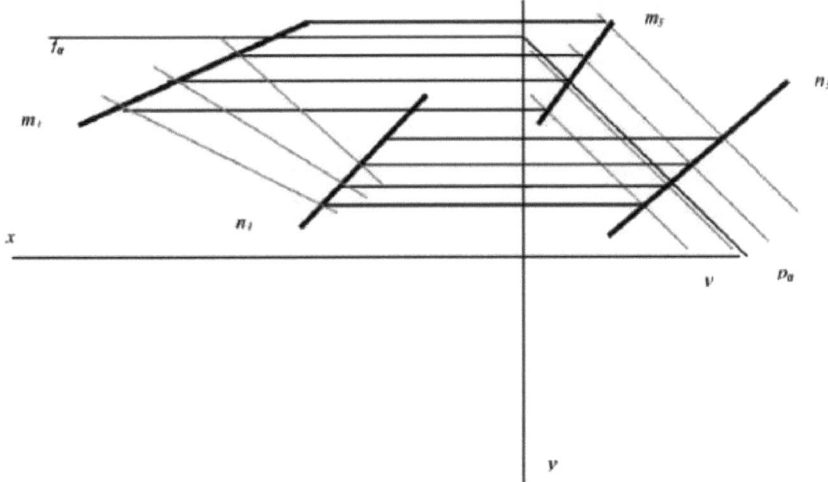

Fig. 4.9

If as the any one guide is taken a plane, the angle between generators and this plane is taken constant (φ_0=const), then is received a surface with the guide plane. If φ_0 is zero, i.e. the generators are located parallel to the guide plane (in this case the guide plane is called the plane of parallelism), is received the surface with the plane of parallelism. There are three types of surfaces: cylindroids, conoid, "aslant plane". Algorithmic part of the determinant is the same for all, only the geometric parts (the guides) are different:

  a) for the cylindroids they are the curves m and n (Fig. 4.7)

  b) for the conoid m is curve , n is straight line (Fig. 4.8)

  c) for " aslant plane" m and n are straights (Fig. 4.9) (on the picture $\alpha =$ π_2).

There is another name for this surface, it is " hyperbolic paraboloid" , because a line of intersection of the surface with a plane is or a parabola , or hyperbola.

4.5 Surfaces of rotation

If the generator of surface (line l) is rotated around straight i (it is called an axis of rotation), is received the surface of rotation. The geometric part of the determinant of surface consists of i and l, and the algorithmic part is condition of rotation the l around the i.

The rotation surfaces are used widely in the technique. This is the consequence of the prevalence of the rotational movement and of the simplicity of processing of rotation surfaces on the machines. Consider a few surfaces of rotation:

  a) Thor. It is formed with the rotating a circumference around an axis. The rotation axis lies in the plane of the circumference, but doesn't pass through its center. Thor can be opened (Fig. 4.10,a) or closed (Fig. 4.10, b) ;

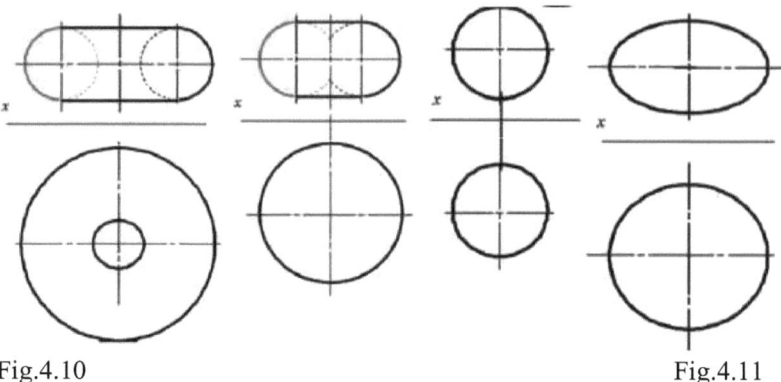

Fig.4.10   Fig.4.11

b) Sphere. It is formed with the rotating a circumference around any diameter (Fig. 4.11,a). Line of section of the sphere with a plane is the circumference.

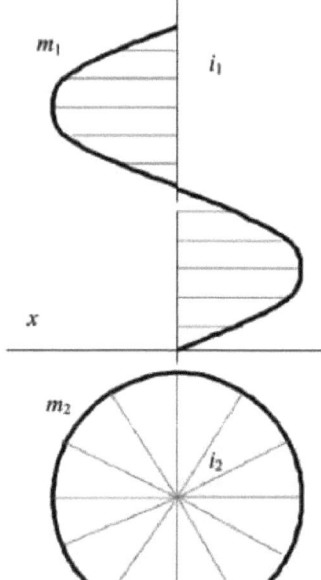

c) Ellipsoid of rotation. It is formed with the rotating an ellipse around its axis. If an axis of rotation is the minor axis, is received a compressed ellipsoid (Fig. 4.11, b), and if the axis of rotation is the major axis, is received an extended ellipsoid of rotation.

d) A hyperboloid of rotation and a paraboloid of rotation are formed with the rotating the hyperbole or parabola around their axes. The called surfaces are not linear surfaces of rotation. Linear surfaces of rotation are known from school, they are a surface of cone of rotation and a surface of cylinder of rotation.

4.6 Screw surface

Surface is called screw surface, if it is formed in consequence of a screw movement of generator (line l). Geometric part of the determinant of surface consists of the line l, the axis of rotation – the straight i and a step of screw movement. In the algorithmic part it is written: "line l performs the screw movement with definite regularities". If generator of the screw surface is straight lines, then the surface is called the helicoid.

Fig .4.12

If the generator intersects the rotation axis, then the surface is called closed, and if it not intersects, then is called opened. The helicoids are divided depending on a situation of the generator (perpendicularly or not perpendicularly to an axis of rotation) into right and inclined. Fig . 4.12 shows the right closed helicoid (this surface can be considered as the conoid, because screw line m and straight line i are guides and π_2 can be reseived as the plane of parallelism) .

4.7 Intersection of the curve surface with the plane

4.7.1 A general concept of the intersection line construction of a surface with the plane

To construct the line of intersection of a surface with a plane in the general case it is necessary to use the auxiliary secant planes. In the time of an intersection these planes with the given plane and surface the formed lines are intersected mutually as lying on one plane. Then they give a number of points belonging to the sought-for line. Obviously, that the place of secant planes it is necessary to define in according to the condition: the received lines should be intersected mutually. To ensure this condition on the epure it is necessary to find *the exception (datum) points of the intersection line* at first. These points are:
- the projections of the points which separate the visible and invisible parts of the intersection line;
- the projections of the points most and least remote from the projections planes.

The construction will be easy, if auxiliary intermediary plane intersects the surface along simple form lines (straight lines and circumferences).

For constructing the line of intersection of the curved surface with the plane one can use the private ways instead of the described above way:
- if the surface is linear, it is necessary to find the intersection points the generators (straight lines) with the plane, and to connect them in a certain order. In this case, the task is reduced to a task of finding of a point of intersection the straight line with the plane;
- if a form of projection of line is known beforehand, its geometric properties can be used for construction. For example, is known that the line of intersection is ellipse. For constructing of ellipse one can define its major and minor axes and use the way known from a course of geometric constructions.

4.7.2 The construction of an intersection line of the surface with a plane with the way of the auxiliary secant planes

Consider an example of using of the auxiliary secant planes way. Let's necessary to construct the projections of the intersection line of the right cone surface with the plane α (Fig. 4.13). In order to construct the projections of the intersection line can use the auxiliary secant planes, such simple lines one can receive onto the surface:
- the straights (if the intersecting planes is passed through the vertex of the cone);
- the circumferences (if the intersecting planes is parallel to a base of the cone). The exception points of the intersection line are found following:
a) find the points most and least remote from the plane π_2 with use of the plane β passing through the cone axis perpendicularly to plane α: $(HF) \cap (S1) = A$, $(HF) \cap (S2) = B$;

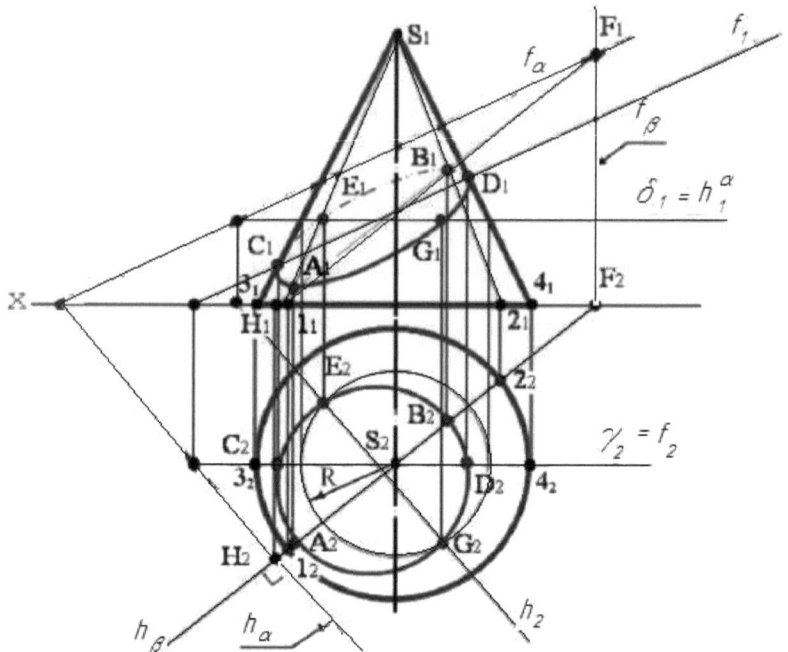

Fig. 4.13

b) to find the point separating the visible and invisible parts of the frontal projection of intersection line with use of the frontal plane γ passing through the axis of the cone : $\gamma \cap (S3) = C$, $\gamma \cap (S4) = D$.

Now it is necessary to draw the plane δ located above the point A and below the point B. This plane intersects the surface of the cone along the circumference (O, R), the plane α – along the contour ($\delta_1 = h_{\alpha 1}$), and the mutual intersection of these lines gets two points: (O,R) $\cap$ $h = E$, G. The received points are connected in a certain order.

4.8 Mutual intersection of surfaces
4.8.1 The general notion about the construction of a line of mutual intersection of two surfaces

I order to construct the line of mutual intersection of two surfaces is used a universal way: the several points of this line are found with the intermediary plane or the intermediary surface (Fig. 4.14). Conical surface, cylindrical surface or a sphere can be taken as the intermediary surface. The intermediary plane may be general position and projected.

The place of the intermediary planes or intermediary surfaces is determined after of finding of the exception points of the intersection line. In some cases the special ways one can use in depending from a type of surfaces and mutual their situation:

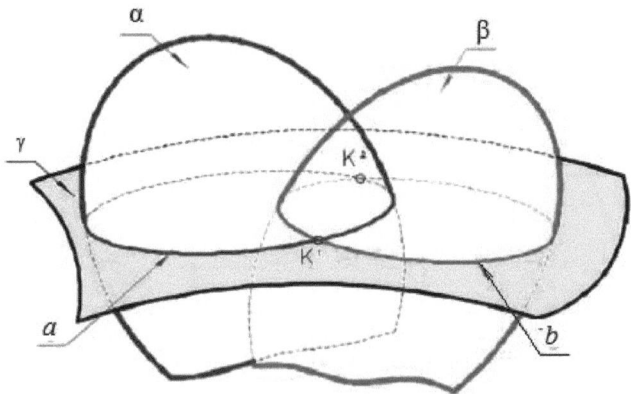

Fig. 4.14

- if one of surfaces is linear surface, one can find the points of intersection of several generators (straight lines) of this surface with the second surface, and connect them in a certain order;
- if the surfaces are face surfaces, it is necessary to solve the tasks on the mutual intersection of two planes or line and plane several times;
- if is known a form of projection of intersection line in advance, one can use its geometric properties. In this case the constructions are become easier and accurate.

4.8.2 Use of auxiliary secant planes

As mentioned above, auxiliary secant plane can be the plane of general or special position. In order to select their place it is necessary to perform one condition: the plane must intersect the given surfaces along the lines of the simple form (straight or circumference). For example, if for constructing of the mutual intersection line of two conical surfaces the plane passing through the vertices of these surfaces to use, then the planes will intersect the given surfaces along generators (straight lines). For constructing of an intersection line of two cylindrical surfaces it is necessary to use secant planes which are parallel to the generators of these surfaces, because the planes intersect both surfaces along straight lines. Now think over, how need draw the planes for constructing the intersection line of conical and cylindrical surfaces.

For constructing the line of mutual intersection of surface of the rotation cone and sphere (Fig. 4.15), one can use the horizontal planes, because they intersect both surfaces along circumferences. In order to select the place of planes it is necessary to find the points most and least remote from the plane π_2 (exception points of intersection line) at first. These points lie on the projections of outlining generators on the plane π_1. In the picture they are found with help of the plane γ: these are points A and B received as a result of the intersection of the median section line of the sphere and generator $1S(1_1S_1, 1_2S_2)$ of the cone surface.

51

Now one can find the intermediate points of the intersection line. Between points A and B a few horizontal planes are drawn. These planes intersect the sphere

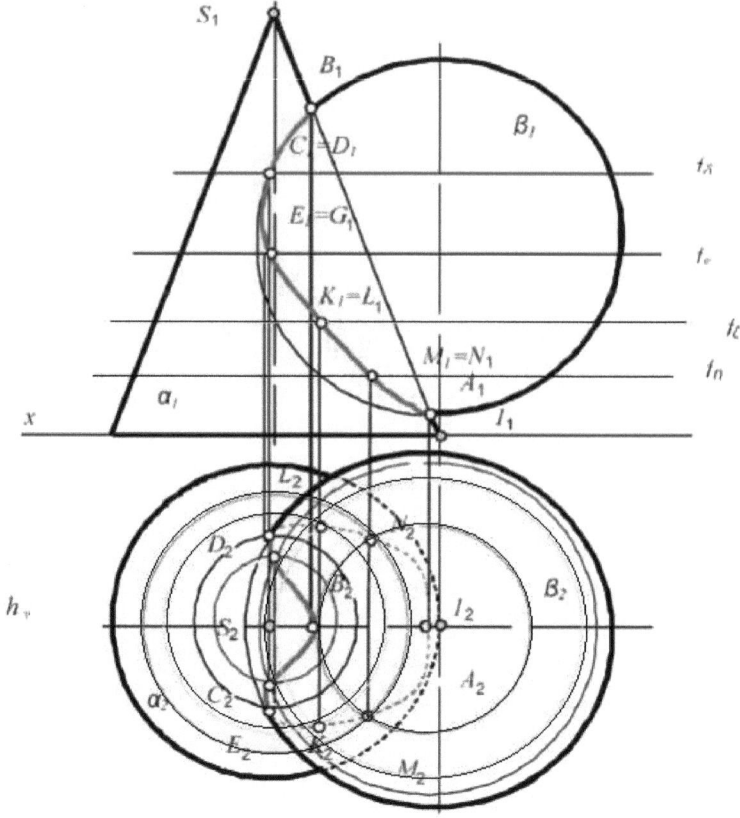

Fig.4.15

and the conical surface along circumferences. These circumferences are projected on the projections plane π_2 without distortion. A pair of circumferences lying on the auxiliary planes gives the points C, ..., N. Now it is necessary to connect their projections with the fair lines. Further it is necessary to noted that with the help of the plane ε passing through the sphere center are found the points E and G of the intersection line which separate the visible and invisible part of the intersection line on the horizontal projection plane.

4.8.3 Use of auxiliary secant spheres

The surfaces must have a common plane of symmetry for using of secant spheres. Spheres can be concentric or eccentric.

Way of eccentric spheres is used in cases when the surfaces are the rotation surfaces, but axis of rotation are crossed, or when one surface is the surface of

52

rotation and second has a parallel circular section (secant plane must not be perpendicularly to axis of rotation surface).

Before considering of way of concentric spheres it is necessary to stop on a special case of intersection of rotation surfaces.

It is known that a surface of a circular cylinder is formed with rotation a straight around of axis parallel to it, and a surface of a circular cone is formed with

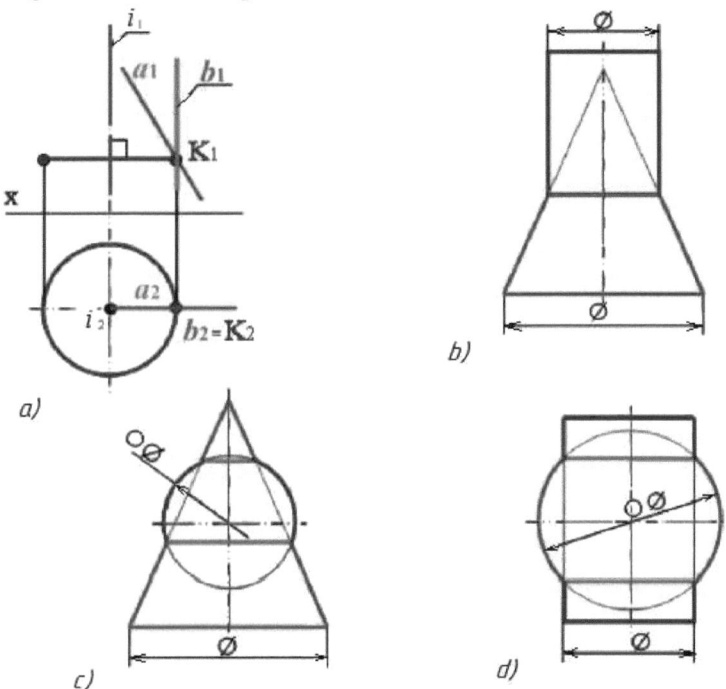

Fig. 4.16

the rotation a straight around axis its intersecting.

Let the axes of these surfaces are coincided. Fig. 4.16 shows the axis of rotation i (it is a horizontally-projecting straight) and generators of these surfaces are straight lines a and b (a and b are coplanar straight, because they are located in one plane). For formatting of the rotation surfaces all points when lie on lines a and b (except the points of an intersection of a and i) describe the circumferences. However, the overall circumference for two surfaces is described only a point of intersection of the two lines a and b – it is a point K. So, one can make such a conclusion: two rotation surfaces with the common axis are intersected on circumferences, and a number of circumferences will be equal to a number of intersection points of coplanar generators located on one side of the rotation axis. Fig. 4.16,b,c,d shows examples of proving this theorem.

Note: only one projection of surfaces are shown on Fig. 4.16, b, c, d , and signs of diameter (∅) or sphere (○) is used in order the image was the reversible image.

The considered case lies on the basis of way of concentric spheres.

Two conditions must fulfill for using of way
- both surfaces must be the rotation surfaces;
- axis of surfaces must intersect, and if they are parallel to any projections plane, the constructions are become easier.

The centers of the spheres must lie in the point of intersection of the axes. For selecting of the radius of spheres it is necessary to find the great radius

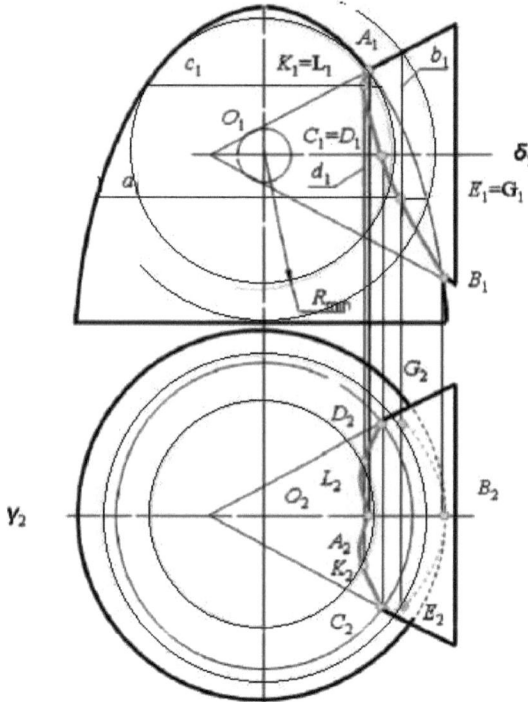

Fig. 4.17

R_{max} and small radius R_{min}. R_{min} is equal to the radius of the large sphere tangency to the surfaces, and R_{max} is equal to the distance from the intersection point of an axes of surfaces to the exception point of intersection line, which is farther point from the intersection point of the axes. Way of spheres is based on the fact that the circumference is formed as a result of the intersection a sphere with the rotation surface.

Consider the use of the way of spheres. It is necessary to construct an intersection line of the surfaces of a rotation solid and the rotation cone (Fig. 4.17). In this case cannot use the way of secant planes, because there are only two planes which intersect both surfaces on simple lines. The use of the way of spheres is very effective, because both the conditions are fulfilled, and the axes of surfaces are parallel to one of the projection planes (π_1). Find the exception points of intersection line at first. As the symmetry plane is general plane for surfaces and is parallel to the plane π_1, with the help of this plane one can define two exception points – the points which separate the visible and invisible parts of the frontal projection of intersection line (in the picture this is the plane γ and the points A, B). These points will be points most and least remote from the plane π_2. The points which separate the visible and invisible parts of the horizontal projection of the intersection line are found with help the horizontal plane δ passing through an axis of the cone. These are the points C and D.

For determining the radii of auxiliary secant spheres it is necessary to find R_{max}, R_{min}. Among previously found points A, B, C, D the point B is the farthermost point from O – the intersection point of the axes of surfaces, therefore $R_{max} = [OB]$. For finding Rmin it is necessary to draw two spheres tangency to the surfaces, with centers into the point O. The radius of most of them will be equal R_{min}. Now one can draw the arbitrary spheres with radius $R_{min} \leq R \angle R_{max}$. Frontal projection of this sphere is shown in the picture. The sphere intersects two given surface along circumferences. These circumferences are projected on the plane π_1 as segments [a_1] and [b_1]. Mutual intersecting of these segments is given the projections of points E and G lying at the intersection line. The points E_2 and G_2 are defined with the help of tie line and circumference (O_2, R_1).

Two points of the intersection line are defined with a help of the sphere with radius is equal R_{min}. For this it is necessary to use the circumferences c and d. Found projections of points are connected with fair lines.

Questions and Exercises

1. What is the kinematic way of the forming of a surface?
2 . What is a framework of a surface?
3 . What is the determinant of a surface?
4 . What is an outline of a surface?
5 . Give the general scheme of a classification of the surfaces.
6. How is a surface of rotation formed?
7. How is a screw surface formed?
8. Describe the algorithm of solving of task of constructing a line of surfaces intersection.
9. What conditions must be fulfill for using of the way of spheres?
10 . Which points of the line of intersection a plane with surface or the two surfaces are called the exception points?
11. How many stages have the solution of the task of finding a point of intersection a straight line with a plane?
12. In what cases the plane intersects the surface of a right circular cone along the straight lines (along a circle, a parabola, a hyperbola, an ellipse)?

13. How many stages have a solution of the task of finding a point of intersection a straight line with a surface?

14. What it is necessary to know for selecting of auxiliary surfaces (planes)?

15. Construct a standard aslant-angular frontal dimetric projection of the curve *l* (Fig.1).

16. Construct a standard rectangular isometric projection of right circular cone with height of 100 and a base diameter of 60.

17. Finish the horizontal projection of the pyramid with a cutout and construct its rectangular dimetric (Fig. 2) .

18. Finish the horizontal projection of a right circular cone with a cutout and construct its rectangular isometric (Fig. 3) .

19. Construct the line of intersection of the sphere with the cone surface with way of the sphere (Fig. 4).

20. Construct the line of intersection of the surfaces of cylinder and cone (Fig. 5)

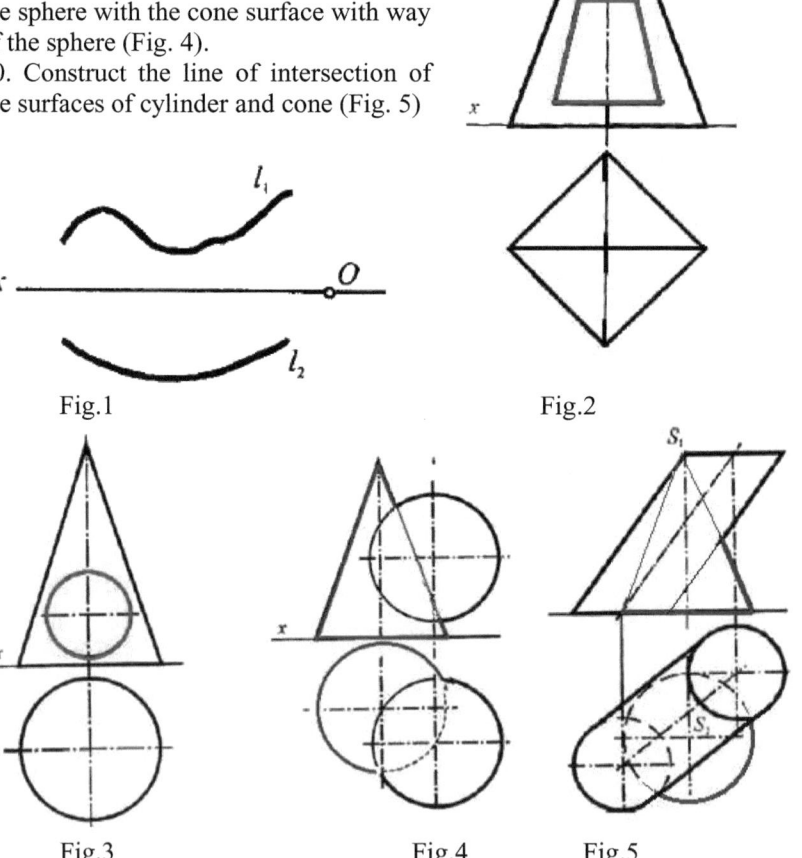

Fig.1 Fig.2

Fig.3 Fig.4 Fig.5

Chapter 5. DEVELOPMENT OF SURFACES

5.1 A general concept about developments

The construction of the developments is one of the important problems of the technique, because on the production different structures which consist from the tanks and pipes made from thin sheets is used often. In order to design these structures it is necessary to construct the developments.

A geometrical aspect the constructing of development of the surface is understood as the transformation of the points of the surface in the points of specially selected plane. Thus, if the surface and development is considered as sets of points, then unambiguous correspondence must be between there. This means that only one point on the development corresponds to every point of the surface and, on the contrary, single point of the surface corresponds to each point of the development.

5.2 Constructing a development of faceted surfaces
5.2.1 Ways of constructing a development of faceted surfaces

Development of faceted surface is a figure composing of its faces. Therefore the life-size of all faces is defined for constructing the development at first. The faces of surface it is necessary to divide into triangles, to find the life-sizes of the sides of the triangles and to locate them in a certain order. It is the easiest way of solving this task. This way is a universal way used for constructing the developments. Along with this there is a way used only for constructing the development of prismatic surface. It is way of normal section. The essence of the way consists in following: for receiving the normal section it is necessary to make:
- draw the plane perpendicularly to the edges of the prism;
- develop the intersection line in a straight line;
- draw straight perpendicularly to the developmental line from the corresponding points;
- measure off and put the life-sizes of the prism edges on these perpendiculars.

Then the place of the prism verities is defined on the development. If to connect the received points, the broken line limiting the development is formed.

5.2.2 Examples of constructing the developments of surfaces

Example 1. Construct the line of intersection of the pyramid surface with the plane α and show this line on development of surfaces (Fig. 5.1). The figure of section is a triangle; from the picture it one can see clearly. As the section plane is frontal-projecting plane, for defining of the intersection line it is necessary to find the frontal projections of vertices of the triangle KLM at first, and then used the tie lines to construct the horizontal projection.

The development of the pyramid surface consists of triangles. Each triangle is equal to the life-size of pyramid face. On Fig. 5.2 one of these triangles is constructed. It is the triangle 2*S*3*, equal to the life-size of face 2S3 of pyramid. For constructing of this triangle it is necessary to realize the auxiliary construction,

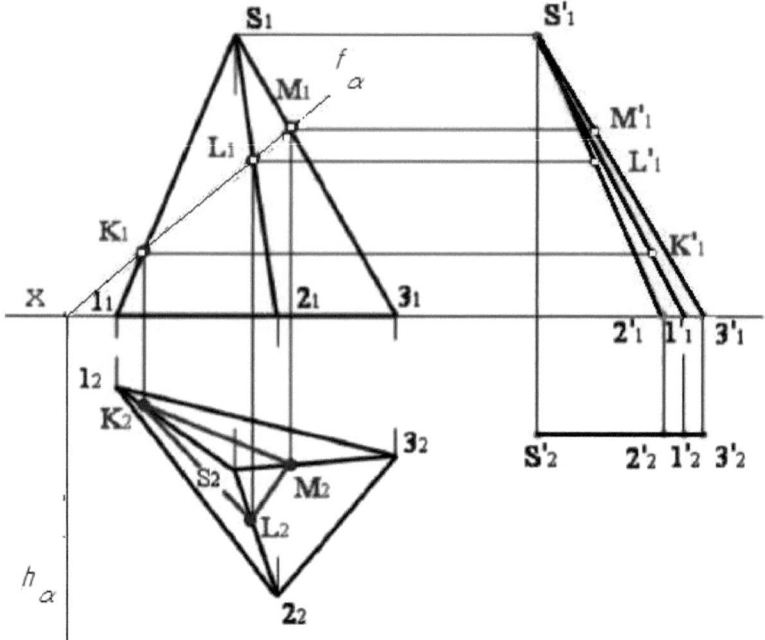

Fig.5.1

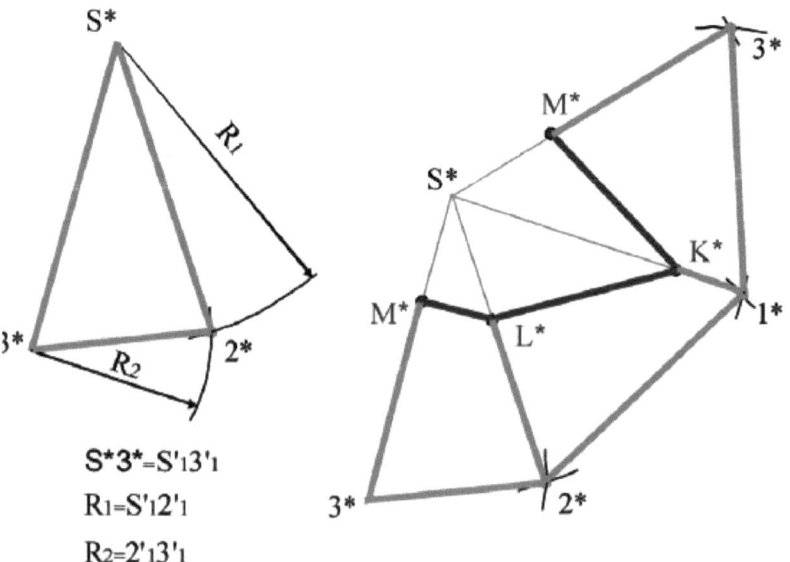

S*3*=S'₁3'₁
R₁=S'₁2'₁
R₂=2'₁3'₁

Fig.5.2

because on the epure of the given pyramid there is no life-size of the lateral edges. On the epure only life-sizes of the pyramid base sides can be found, because the base is located on the plane π_2: $[1_2 2_2] = |12|$, $[2_2 3_2] = |23|$, $[3_2 1_2] = |31|$. In the result of such constructions the horizontal projections of the pyramid edges $S1$, $S2$, $S3$ will be located parallel to the x-axis and their frontal projections will be equal to the life-sizes of edges: $[S'_1 1'_1] = |S1|$, $[S'_1 2'_1] = |S2|$, $[S'_1 3'_1] = |S3|$. In this case, the way of plat-parallel movement is used. Together with the edges of the pyramid the points K, L, M lying on them are moved, therefore $[S'_1 K'_1]$, $[S'_1 L'_1]$, $[S'_1 M'_1]$ will be equal to the life-sizes of the corresponding segments.

Shown on fig. 5.2 triangle $S*2*3*$ is received with such way: segment $[S*3*]$ = $|S3|$ is constructed on the free place, then two arcs are drawn: one – from the point $S*$ with radius $R_1 = [S2]$, a second – from point $3*$ with radius $R_2 = [32]$. The point $2*$ lies in the intersection of these arcs. The triangles $2*S*1*$ and $1*S*3*$ are constructed with similar means, the resulting figure will be the development of the lateral surface of the pyramid. For constructing of the line of intersection on the development it is necessary to find the places of points $K*$, $L*$, $M*$: $[S*K*] = [S_1 K_1]$, $[S*L*] = [S_1 L'_1]$, $[S*M*] = [S_1 M'_1]$, then to connect them in a certain order (Fig. 5.2).

Example 2. Construct the development of the surface of inclined prism (fig. 5.3).

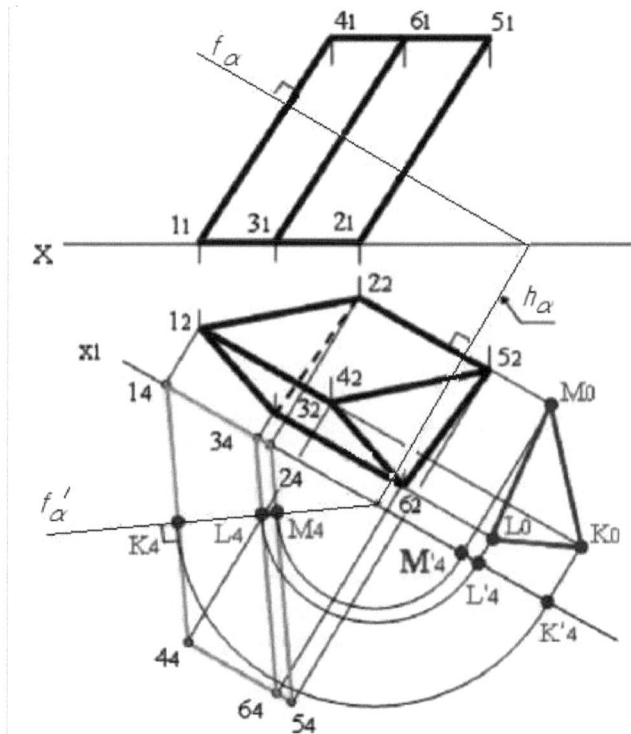

Fig.5.3

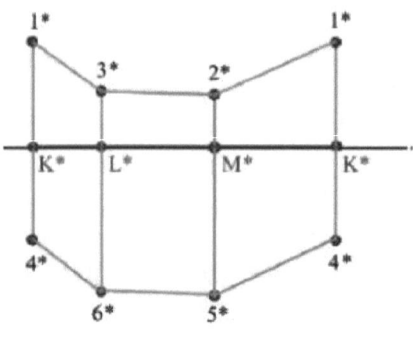

It was said above, for constructing the development of the surface of inclined prism, one can use the way of normal section. Draw the plane α perpendicularly the edges of the prism (normal plane) in an arbitrary place. In the system π_1, π_2 this plane is the plane of general position, therefore, in order to it was projecting plane it is necessary to pass to another system. In the picture the plane π_4 is introduced. In the result of the transition to another system are solved two problems: firstly, the plane α comes in projecting position (figure of section is projected on the

Fig.5.4

plane π_4 as a segment of straight line), secondly, the projections of the prism edges on the plane π_4 will be equal to their life-sizes. Now it is necessary to find the life-size of the figure of section. In the picture task is solved with the way of coincidence of the plane α with the plane π_2: $\Delta KLM = \Delta K_0 L_0 M_0$. After that one can construct the development. Draw the straight on a arbitrary place, then construct in the following order the segments $K*L* = K_0 L_0$, $L*M* = L_0 M_0$, $M*K* = M_0 K_0$. Now it is necessary to construct the straight lines perpendicularly to these lines through the points $K*$, $L*$, M, measure and put the segments $K*1* = K_4 1_4$, $L*3* = L_4 3_4$, $M*2* = M_4 2_4$, $K*4* = K_4 4_4$, $L*6* = L_4 6_4$, $M*5* = M_4 5_4$ on them. And so, the places of the prism vertices on the development are found. The found points is connected in a certain order (Fig. 5.4).

5.3 Construction the developments of the curve surfaces
5.3.1 Ways of constructing the developments of the curve surfaces

Ways of construction the developments of the surface of right circular cone or cylinder are known from school, therefore they is considered not here. Construction of the development of the conical and cylindrical surfaces which are not surfaces of rotation, it is recommended to perform in the following order:

1) inscribe in the surface or describe around the surface the n-faceted surface; a number n depends on a size of the drawing, but in any case, it number will be not less 8;

2) construct the development of the n-faceted surface as shown previously;

3) connect the ends of the edges on the development with the fair lines.

5.3.2 Examples of the construction the developments of the curve surfaces

Fig. 5.5 shows the development of the surface of a right circular cone, fig. 5.6 shows development of the surface of a right circular cylinder constructing with the help of the orthogonal projection of the cone (cylinder). Side by side with this, on pictures the place of point A lying on a surface of the cone (cylinder) is shown on the development.

The developments of surfaces of the cylinder and cone are shown in fig. 5.7, 5.8 and 5.9. Can see from constructing that 8-multifaceted pyramid is inscribed in the conical, and 8-multifaceted prism – in the cylindrical surface.

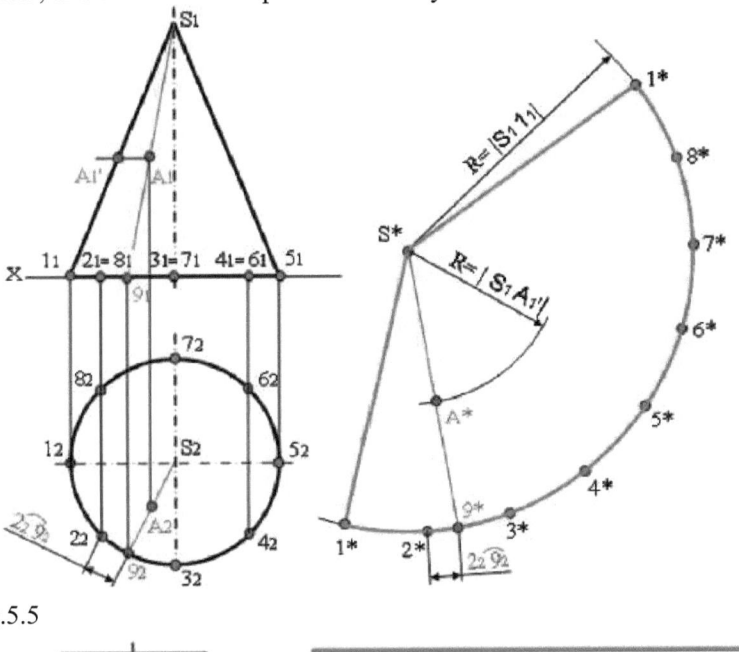

Fig.5.5

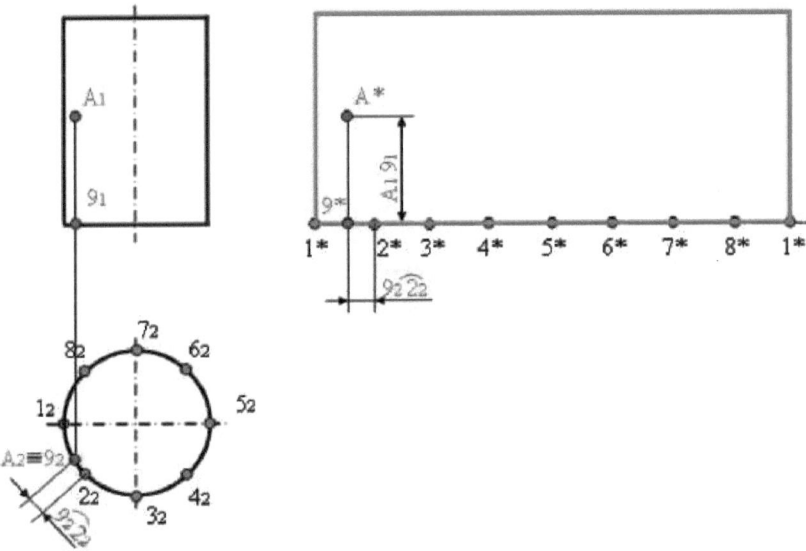

Fig.5.6

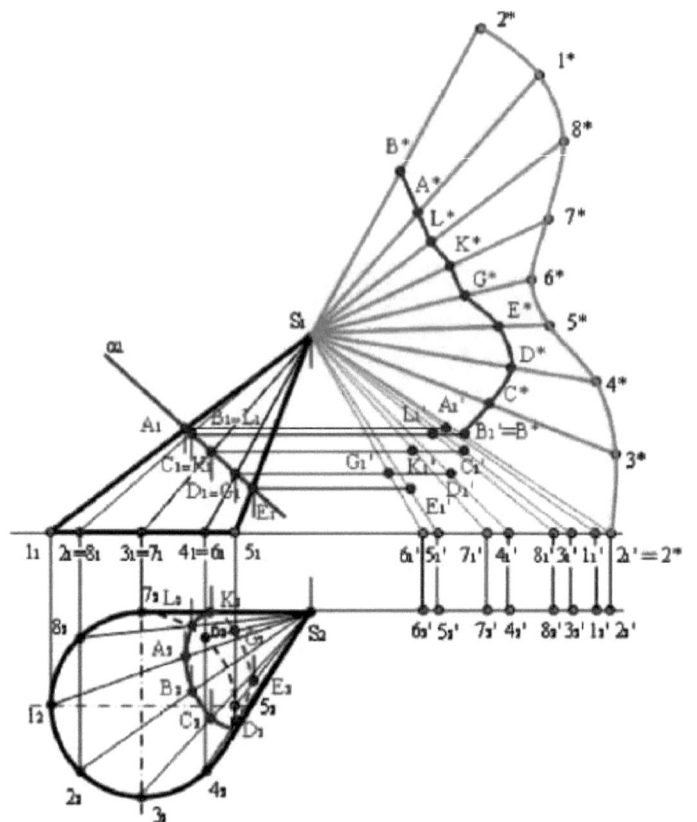

Fig.5.7

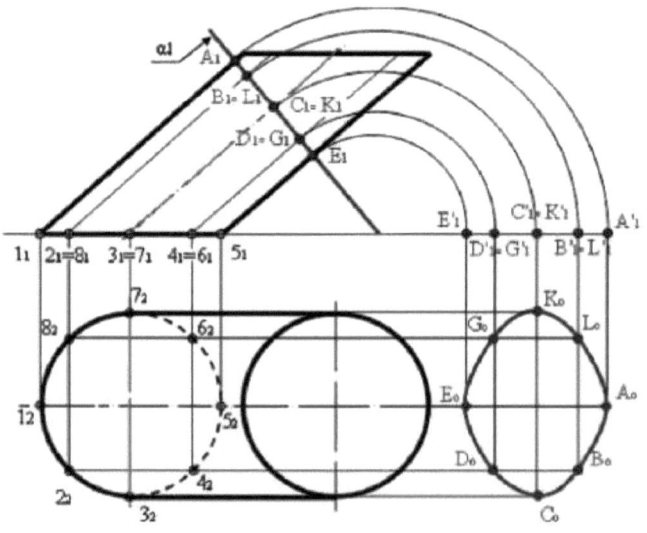

Fig.5.8

62

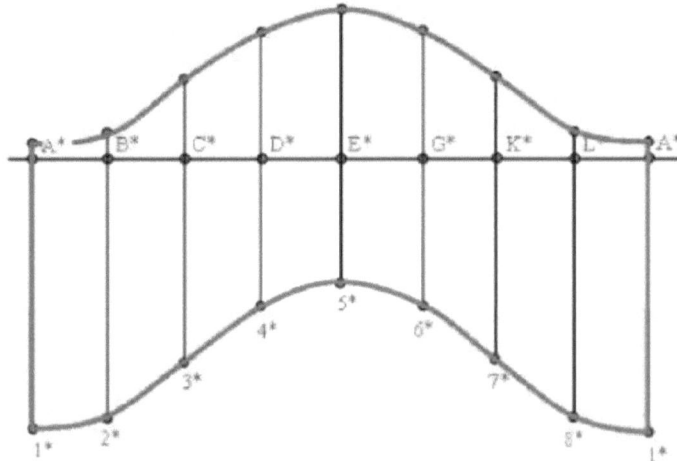

Fig.5.9

Questions and Exercises

1. What is a development of surface?
2. What surfaces are developmental?
3. Call the ways of the constructing of developments. Show, in which cases which way is rational way.
4. For which surfaces can construct the mathematically precise developments?
5. Complete the conditions of tasks 19 and 20, given at the end of Chapter 4, with the words: "construct the developments of given surfaces and mark the intersection line on them" and continue solving the tasks.

Literature

1. Русско-казахско-английский политехнический словарь: более 80000 терминов и словосочетаний. В 2 т.- Т.1 - Алматы: Rond&A, 2010.-740 с. .- Т.2 - Алматы: Rond&A, 2010.-720 с.
2. Colin H.Simmons, Dennis E.Maguire. Manual of Engineering Drawing to British and International Standards: Second edition.-Elsevier, 2005.-298 P.